GEOLOGY *of the* NORTH CASCADES

A Mountain Mosaic

ROWLAND TABOR AND RALPH HAUGERUD

Drawings by Anne Crowder

THE MOUNTAINEERS BOOKS

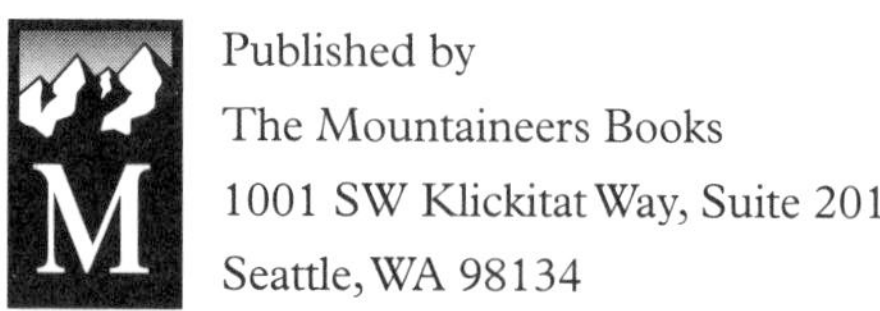

Published by
The Mountaineers Books
1001 SW Klickitat Way, Suite 201
Seattle, WA 98134

First printing 1999, second printing 2002, third printing 2006, fourth printing 2010, fifth printing 2017

Distributed in the United Kingdom by Cordee, www.cordee.co.uk

Manufactured in the United States of America

Edited by Stephen R. Whitney
Drawings by Anne Crowder
Cover and book design by Ani Rucki
Layout by Ani Rucki

Cover photograph: *North Cascades National Park, WA. Dike and Southern Pickets* © James Martin.

Library of Congress Cataloging-in-Publication Data
Tabor, R.W. (Rowland W.)
Geology of the North Cascades : a mountain mosaic / Rowland Tabor and Ralph Haugerud ; drawings by Anne Crowder. — 1st ed.
p. cm.
Includes bibliographical references.
ISBN 0-89886-623-5
1. Geology—Cascade Range. 2. Geology—Washington (State)
I. Haugerud, Ralph Albert. II. title.
QE176.C37T33 1999
557.97'5—dc21 98–52131
CIP

Printed on recycled paper
ISBN (paperback): 978-089886-623-0
ISBN (ebook): 978-1-59485-304-3

Contents

This book is dedicated to everyone who loves

the North Cascade Range and would like to know more about its rocks,

And especially to:

Fred Beckey who climbed all its mountains and wanted to know all about its rocks,

the late Peter Misch who knew all about its rocks and wanted to climb all its mountains,

and the late Dwight Crowder who knew its mountains and its rocks

and would have liked to have written this book.

What This Book Is All About

The deep wilderness and rugged peaks of the North Cascade Range are only a few hours' drive from the metropoli of Washington and British Columbia (Figure 1). The average visitor to the range experiences a bit of Alaska and the Alps for the price of a weekend trip. The spectacular scenery of these mountains is carved from equally spectacular geology. The steep cliffs, horrendous brush, and long trails that guard North Cascade peaks have their geologic counterparts, for an understanding of the geology of the range can be as difficult to gain as any North Cascade summit. We have spent many decades puzzling out some of the geologic history of the range. This book is a guide to what we and others have learned.

As with any mountain excursion, preparation makes for a more enjoyable experience. Part I of this book summarizes the geologic history of the North Cascades and includes a primer on geologic terms and processes. We hope this primer, along with the glossary at the end of the book, helps readers untrained in geology understand the complex and fascinating story told by the rocks of the North Cascades. Readers with some geologic training can skip the primer and delve into the detailed Geologic Notes in Part II. We haven't pulled our geologic punches much, so the more experienced student of geology will also find useful information in Part I of this guide.

Part II comprises a series of Geologic Notes describing various geologic features of interest at locales throughout the North Cascades. The features exemplify general concepts discussed in Part I, where they are cross-referenced as appropriate. They are arranged by major rivers and streams of the range, beginning with the Skagit River in the northwest and ending with the upper Pasayten River in the northeast. Points of geologic interest that are located well off major drainages and reached only by hiking are listed within those drainages offering the most probable approach routes. The location and access route for each Geologic Note are shown on the "Points of Geologic Interest" maps (see Plate 7A–D).

Because many of the ideas about rocks and geologic processes in the North Cascades presented here are relatively new, or at least not commonly known, even within the geologic community, we have included in "Reading and References" an annotated bibliography of the more important scientific references. Many of these sources may not be of interest to the general reader, but their titles give flavor to the history of the geologic ideas. We have also noted background reading of more general interest.

Acknowledgments

To summarize more than 90 years of geologic study, we must draw on the work of many, many scientists. The story of the North Cascades came not only from hard intellectual work, but from a great deal of hard legwork. We are indebted to all the geologists who struggled up brushy creeks, traversed rocky hillsides, and endured rainy days. Some were professional explorers, some teachers, and some students. We thank them for their help. Of the major contributors to North Cascade geology that we have mentioned in the main text of this book, we especially want to thank colleagues and friends: the late Peter Misch who taught us

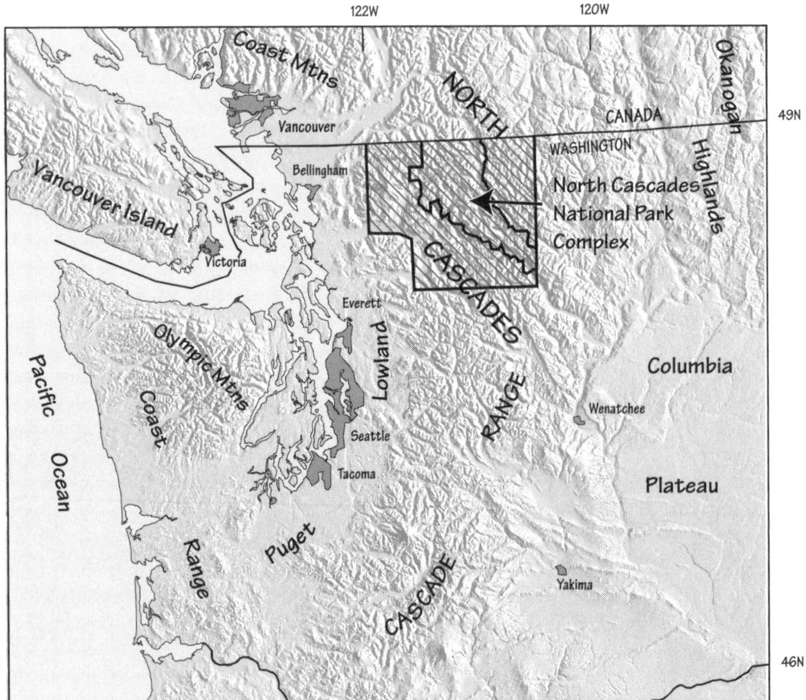

Figure 1. Location of the North Cascades. This guide covers the lined area.

and inspired our love of these mountains, Ned Brown who resolved many jumbles of geologic observations into elegant theories which he has shared enthusiastically with us and his other students, and Joe Vance who, with a mountaineer's passion, has steadfastly wrested geologic secrets from North Cascade terrain over a span of more than 40 years. In addition we have gained much from the work of, and discussions with, Scott Babcock, Derek Booth, Darrel Cowan, Joe Dragovich, Wes Hildreth, Cliff Hopson, Sam Johnson, Mike McGroder, Bob Miller, Jim Monger, Jon Riedel, Jeff Tepper, and Donna Whitney.

Illustrations for this book are from many sources. Most of Anne Crowder's drawings are initialed, but she also drew figures 2, 3, 6, 8, 10, 11, 15, 17, 21, 22, 24, 26, 27, 31, 39, 41, 59, 65, 81, and 100. Figures 3, 4, 5, 12, 18, 29, and 80 are by Anne and the authors. Part of illustration 3 and illustrations 34, 36, 37, 40, 54, 55, 56, 62, 63, 67, 73, 86, 88, 91, 92, 95, 96, 102, 103, and 104 are by Ed Hanson, mostly initialed and attributed to source in the references. Some have been much modified. Illustrators for adopted figures 62, 74, and 109 are identified in the cited material. Esther McDermatt drew figures 90 and 93. All other unattributed illustrations, maps, and photographs are by the authors.

We thank all the people of the National Park Service and the U.S. Forest Service who have helped us study the geology and encouraged our work on this book. In particular, personnel of the Marblemount Ranger Station, Stehekin Visitor Center, and Winthrop Ranger Station have helped in many ways. Special thanks to Kevin Kennedy at the Glacier Visitor Center.

This book has been markedly improved by reviews and endless good suggestions from Melanie Moreno, King Huber, and Wes Hildreth. Derek Booth, Ned Brown, Jon Riedel, and Whitney Tabor also read drafts of the manuscript and made helpful comments. The authors thank Kajsa Tabor for companionship on geologic hikes, photographs, meticulous editing of early drafts, and constant encouragement to explain things better. The nongeologist reader owes much to Kajsa's clarification. The junior author thanks Martha Bean for sharing the joys of both North Cascades geology and parenting, often allowing him to concentrate on the former at the expense of the latter.

And even after all this help, we still needed more. Thanks to our editor, Stephen Whitney, and to the staff at The Mountaineers Books, especially Ani Rucki, Cynthia Newman Bohn, and Margaret Foster, who with great patience helped bring this book into the world.

PART 1

NORTH CASCADES GEOLOGY

A Mountain Story

chapter 1

World Class and Close to Home

The North Cascade Range in Washington State is part of the American Cordillera, a mighty mountain chain stretching more than 12,000 miles from Tierra del Fuego to the Alaskan Peninsula, and second only to the Alpine-Himalayan chain in height and grandeur. Although only a small part of the Cordillera, mile for mile the North Cascade Range is steeper and wetter than most other ranges in the conterminous United States. In alpine scenery and geology, the range has more in common with the coast ranges of British Columbia and Alaska than it does with its Cordilleran cousins in the dry Rocky Mountains or benign Sierra Nevada. Although the peaks of the North Cascades do not reach great elevations (high peaks are generally in the 7,000- to-8,000-foot range), their overall relief, that is, the relatively uninterrupted vertical distance from valley bottom to mountain top, is commonly 4,000 to 6,000 feet, a respectable height in any world-class mountain range. Much of the range is roadless wilderness preserved from commercial exploitation by inclusion in North Cascades National Park, the Ross Lake and Lake Chelan National Recreational Areas (both managed by the National Park Service), and several dedicated wilderness areas managed by the U.S. Forest Service.

Catching the Clouds

Because North Cascade peaks intercept frequent water-laden storms sweeping in from the Pacific Ocean, the range receives large amounts of rain and snow. This constant watering supports luxuriant vegetation on the west side of the range and a swarm of glaciers and snowfields on the higher peaks. From west to east the amount of precipitation changes greatly: some western valleys receive up to 160 inches of precipitation a year, whereas east of the Cascade crest, precipitation is only 10 to 20 inches (Figure 2).

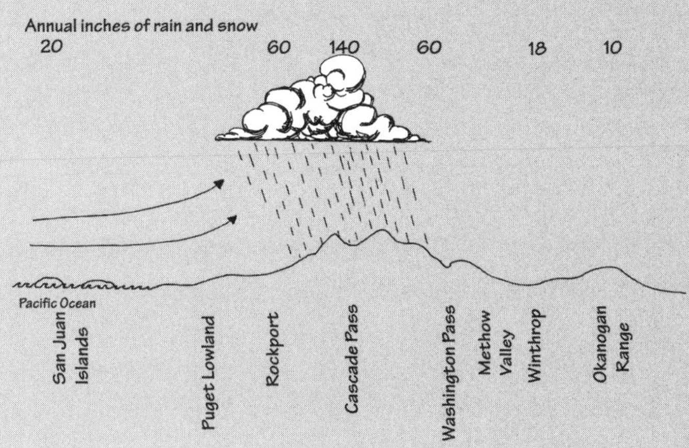

Figure 2. Amounts of rain and snowfall from west to east across the North Cascades. As warm, moisture-laden air from the ocean rises over the mountains, it cools and can no longer hold so much water. It drops most of it on the west side and on the crest of the mountains as rain and snow.

Rocks of the North Cascades record at least 400 million years in the history of this restless Earth—time enough to have collected a jumble of different rocks. The range is a geologic mosaic made up of volcanic island arcs, deep ocean sediments, basaltic ocean floor, parts of old continents, submarine fans, and even pieces of the deep subcrustal mantle of the earth. The disparate pieces of the North Cascade mosaic were born far from one another but subsequently drifted together, carried along by the ever-moving tectonic plates that make up the Earth's outer shell. Over time, the moving plates eventually beached the various pieces of the mosaic at a place we now call western Washington.

As if this mosaic of unrelated pieces were not complex enough, some of the assembled pieces were uplifted, eroded by streams, and then locally buried in their own eroded debris; other pieces were forced deep into the Earth to be heated and squeezed, almost beyond recognition, and then raised again to our view.

About 40 million years ago a volcanic arc grew across this complex mosaic of old terranes. Volcanoes erupted to cover the older rocks with lava and ash. Large masses of molten rock invaded the older rocks from below. The volcanic arc is still active today, decorating the skyline with the cones of Mount Baker and Glacier Peak.

The deep canyons and sharp peaks of today's North Cascade scene are products of profound erosion. Running water has etched out the grain of the range, landslides have softened the abrupt edges, homegrown glaciers have scoured the peaks and high valleys and, during the Ice Age, the Cordilleran Ice Sheet overrode almost all the range and rearranged courses of streams. Erosion has written, and still writes, its own history in the mountains, but it has also revealed the complex mosaic of the bedrock. There is much to be learned about the processes of nature in this special place.

Early Encounters with the Rocks

The first explorers in the North Cascades who had an interest in rocks may well have been the native peoples, quarrying chert and volcanic glass for tools and weapons. According to archaeologists, hunters quarried chert for weapons as early as 8,000 years ago in the Skagit River valley near what is now Ross Lake. Based on radiocarbon ages (see Glossary) of old campfires at quarry sites, archaeologists estimate that the early rock-workers were most active in the mountains between 3,500 and 4,000 years ago.

Native peoples established trails over some passes in the North Cascades. The first Europeans, seeking furs, gold, and trade routes, followed these trails. By the 1890s, mining claims were established in many areas, including in the vicinity of Harts Pass in the Pasayten Wilderness, the upper North Fork of the Nooksack in the Mount Baker Wilderness, and the Cascade Pass area of North Cascades National Park. The first professional geographer to see a significant part of the backcountry was George Gibbs, who in 1849 explored the Skagit, Chilliwack, and Pasayten drainages. His travel was no doubt difficult, but his rather dry account does no justice to the magnificence of the country.

In the early 1900s, as part of a resurvey of the border along the 49th parallel, Canada and the United States sent geologists to study the area along the border. George Otis Smith and Frank Calkins of the U.S. Geological Survey worked on the U. S. side. Although Smith and Calkins made significant contributions, their report pales beside that of their Canadian counterpart, a young professor from Boston named Reginald Daly. Lumberjacks had cleared trees from a 100-foot-wide swath along the border, allowing Daly, usually mounted on a horse and following the tree fellers' trails, to reach terrain never seen before by geologists. His extensive report, published in 1912, was, for more than 40 years, the major source of geologic lore for the North Cascades. However, vast parts of the North Cascades remained unknown geographically and geologically until more recent times.

chapter 2

A Geology Primer for the North Cascades

Before setting out into the rather complex terrain of North Cascade geology, the reader had best be equipped with some basic geologic vocabulary and conceptual tools. This chapter introduces these words and tools and includes briefings on minerals and rocks, geologic time, and the fundamental geologic structure of the North Cascades. The chapter also introduces readers to the theory of plate tectonics. Some understanding of this "unified field theory" of geology makes it possible to place the North Cascades in the big picture of geologic processes that operate on a worldwide scale.

MINERALS AND ROCKS

Fundamental to understanding mountain geology is the nature of minerals and rocks, for the mountains are made out of rocks, and the rocks are made out of minerals. The simplest, most widely known and distributed mineral is quartz (silicon dioxide), which is made up of the elements silicon and oxygen. Most of the rocks in the North Cascades, and elsewhere for that matter, are made up of about seven types of silicate minerals, all of which contain a dose of silica (as silicon tetroxide), combined with various elements. (See "Common Rock-Forming Minerals and Their Components.")

Rocks are classified according to the ways they are made and the proportions of various minerals that form them. Many rock names reflect their composition.

TYPES OF ROCKS

All rocks belong to one of three major groups: igneous, sedimentary, or metamorphic. Many different rocks representing these three groups occur in the North Cascades.

Igneous Rocks

Igneous rocks form from melted rock, called *magma*. On the surface of the Earth, igneous rocks usually occur in a cooled state, except where molten rock is erupting from an active volcano. Magma that flows out onto the Earth's surface is called *lava*. Rocks that crystallize quickly from cooling lava at or near the Earth's surface are called *volcanic* rocks—no surprise. Black, dense, very fine-grained or even glassy, *basalt* is the most common volcanic rock and well represented in the North Cascades. A close relative of basalt, *andesite,* is also a widespread volcanic rock in the North Cascades. Commonly, angular chunks and pieces of volcanic rock debris are cemented together to form *volcanic breccia.* Volcanic breccia may form from broken pieces of rock blasted out of a volcano or from rubble eroded from it.

Rocks that crystallize from magma very slowly and at great depths in the Earth's crust are called *plutonic* rocks (Figure 3). Pluto was the Greek god of the subterranean world of the dead. Granite, with its sparkling and easily noted crystal faces, is a plutonic rock. Each variety of plutonic rock has its volcanic counterpart. For example, a coarse-grained plutonic granite is chemically equivalent to a fine-grained or glassy volcanic rhyolite (Chapter 6). Because the magma which forms plutonic igneous rocks generally rises into or penetrates existing rocks, it is commonly said to be intrusive, and the rock solidified from it is an intrusive igneous rock.

Geologists often refer to a variety of light-colored plutonic igneous rocks as *granitic* because they look like granite. Granitic rocks, especially the varieties *tonalite* and *granodiorite,* make up a large part of the North Cascades, and, for that matter, the cores of many major mountain ranges.

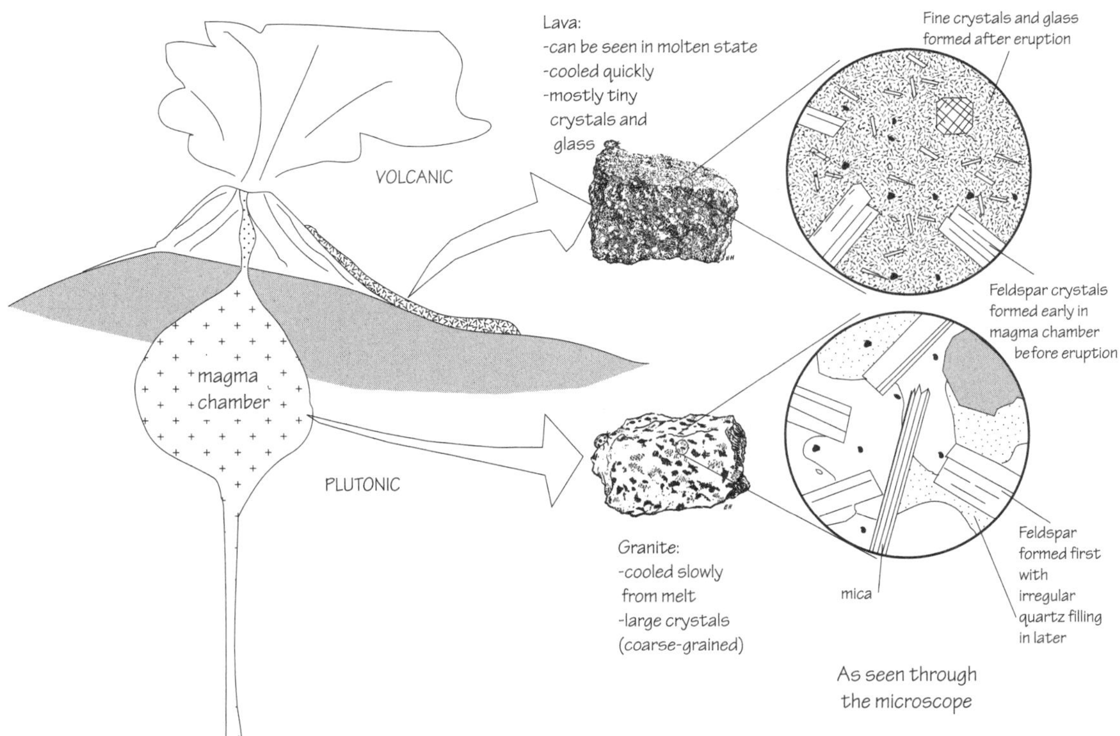

Figure 3. Cross section showing melted rock in magma chamber erupting to form volcanic rock. When the magma crystallizes at depth, it becomes plutonic rock.

For more information on the names of igneous rocks in the North Cascades, see Geologic Notes 93 and 95. Also see "granitic rocks" in the Glossary.

An individual body of granitic rock, solidified from one batch of magma, is called a *pluton.* Very large continuous areas of granitic rocks, commonly made of many plutons, are called *batholiths,* which literally means "deeply formed rocks" (Plate 5A).

Ultramafic rocks are of relatively small extent in the North Cascades but are geologically significant because they are derived from the Earth's mantle (see "Plate Tectonics and Tectonic Terranes" in this chapter). Ultramafic rocks are dark, heavy, and rich in iron and magnesium. They initially formed from melted rock when the Earth first solidified, billions of years ago. They probably have undergone many transformations during their long residence in their deep earth abode, such as metamorphic recrystallization at very high temperatures (see below) and remelting.

Because ultramafic rocks are generally found in relatively small masses—fist- to house-sized—in a variety of other rocks, geologists used to think that the ultramafics invaded the other rocks as magma, that is, that they were intrusive igneous rocks. But throughout much of geologic time, the crust of the Earth was too cold for the high-temperature minerals of ultramafic masses to reach the surface in a melted state. Ultramafic rocks reach the surface of the earth only by way of complex crustal upheavals or as small pieces carried up by other kinds of upwelling magma. By the time most ultramafic materials reach the Earth's surface, they are commonly metamorphosed at lower temperatures to green *serpentine* minerals, which are iron- and magnesium-rich silicate minerals with water bound up in the crystal lattice. A rock made of serpentine minerals is called *serpentinite.* Two especially large masses of ultramafic rocks crop out in the North Cascades: the Twin Sisters massif southwest of Mount Baker and an outcrop in the vicinity of Mount Stuart, south of the area covered in this book. Small pieces of ultramafic rock occur in other areas, and we will describe their significance in subsequent chapters.

Sedimentary Rocks

Debris carried off the continents and into the oceans by streams and rivers is the most common material that

Common Rock-Forming Minerals and Their Components

Mineral	Composition
Light-Colored Silicates	
Quartz	silicon dioxide or silica
Feldspar	silicon tetroxide plus varying amounts of aluminum, sodium, calcium, and potassium
Dark-Colored Silicates	
Mica	silicon tetroxide plus varying amounts of aluminum, potassium, iron, magnesium, calcium, sodium, and water
Pyroxene	silicon tetroxide plus varying amounts of aluminum, iron, magnesium, calcium, and sodium
Amphibole	silicon tetroxide plus varying amounts of aluminum, iron, magnesium, calcium, sodium and water
Olivine	silicon tetroxide plus varying amounts of iron and magnesium
Nonsilicates	
Calcite	calcium carbonate

Of the seven abundant minerals, feldspars, micas, pyroxenes, amphiboles, and olivines are really mineral families, with a variety of named members. We mention other important mineral families as we go along.

Minerals are crystalline substances, meaning that the atoms of their constituent elements are arranged in a definite geometric structure. This structure gives minerals specific physical properties, which geologists, and everyone else for that matter, can identify in the field. We describe these properties as we go along, but for more specific descriptions of the minerals see the Glossary at the end of this book.

makes up sedimentary rocks. Mud becomes *mudstone* or *shale,* sand becomes *sandstone,* and gravel becomes *conglomerate* (evidently the term "gravelstone" never caught on) (Plate 5B). Some sedimentary rocks, however, are not of continental origin but are mostly derived from the skeletons of ocean animals. Accumulations of sea shells or the bodies of tiny calcareous plankton become *limestone,* made up of the mineral calcite (calcium carbonate). Siliceous plankton (known as *radiolarians*) become *chert,* a significant North Cascade rock made up of the mineral quartz. Plant debris becomes coal. Sediments become rock when buried, subjected to pressure, and cemented by mineral precipitates. Geologists, somewhat casually, refer to any hard, dark fine-grained sedimentary rock like shale or mudstone as *argillite* (Plate 4A). Parts of the North Cascades display almost all the named varieties of sedimentary rocks.

Metamorphic Rocks

Metamorphic rocks are changed rocks, that is, rocks whose original materials have recrystallized to form new minerals. Generally, this recrystallization, called *metamorphism,* takes place when deeply buried rocks are subjected to great pressure and high temperature. Most metamorphic rocks have also been squeezed so that their shapes—and the shapes of their constituent minerals—have changed during metamorphism, resulting in a layered or streaky appearance. Metamorphic rocks in the North Cascades formed from preexisting rocks of every kind and from all of the major rock groups, igneous, sedimentary, and metamorphic. For example, *mica schist,* a common metamorphic rock in the North Cascades, recrystallized from shale. *Gneiss* looks like granite which has been squeezed so that the rock looks streaky. Much North Cascade gneiss formed from granitic rocks in exactly that way. Gneiss also forms from schist when the rock remains hot enough for the metamorphic crystals to grow large. *Marble* is metamorphosed limestone, whose calcite crystals have grown large enough to see without a hand lens.

This brief introduction to *petrology,* or the study of rocks, has touched on only a few of the hundreds of different kinds of rocks found in the North Cascades. We name and describe many of the others as the geologic story unfolds.

A ROCK'S PROGRESS—FROM SHALE TO GNEISS OR FROM BASALT TO AMPHIBOLITE

In an ideal metamorphic sequence of increasing heat and pressure, a shale becomes slate and slate becomes phyllite, then schist, and finally gneiss. A shale, when highly squeezed deep in the Earth, first forms *slate,* as the shale's tiny grains and platy clay minerals slide around under the stress. The rock deforms—that is, bends and flows—and the minerals align themselves in the direction of yielding, or flow. Slate exhibits *cleavage,* meaning it can be broken into even, flat chips, slabs, or blackboards (at least in the past). If slate minerals further recrystallize, giving the rocks a shiny appearance, the new rock is *phyllite.* Phyllite also abounds in the North Cascades.

Still further recrystallization produces a flaky crystalline rock called *schist,* usually mica schist made up of flat mica flakes and other minerals. These rocks exhibit *foliation,* a tendency to break into thin, curving leaves, or sheets, like the pages of a book. Thorough recrystallization at higher temperature will produce a gneiss, looking rather like granite, but with minerals clearly arranged in a parallel fashion—the foliation. In gneiss, the minerals feldspar and quartz have grown to become conspicuous crystals. The rock no longer breaks as easily along the foliation because the feldspar and quartz are more randomly oriented and commonly have interlocking, irregular crystal borders.

In progressive metamorphism of a basalt, the course of change is different because the original basalt reacts differently to heat and pressure. In fact, it is so stiff and resistant to the squeezing (unlike wimpy shale) that the first reincarnation as a metamorphic rock is simple recrystallization to a rock called *greenstone*—so named because it is made of many green metamorphic minerals. Further squeezing finally overcomes the basalt or greenstone resistance, forming *greenschist,* which has many of the same minerals as greenstone but with the flaky foliation of all schists. Rising temperature and continued squeezing causes new minerals to crystallize, and what was formerly basalt becomes *amphibolite,* a rock that looks like a dark gneiss and is rich in hornblende and feldspar, but with very little quartz.

Although all the rocks mentioned in these progressions occur in the North Cascades, ambitious hikers will find few places where they can walk continuously from shale or basalt into well-metamorphosed gneiss or amphibolite. In a few places geologists have found parts of the progression, such as phyllite grading into schist, or basalt grading into greenstone.

GEOLOGIC TIME

Even geologists, who have opportunity to practice, have difficulty imagining the great length of time needed for geologic processes. We can think about a million years or even a billion years, but we can hardly imagine the countless small events that fill such expanses of time. And it is just such small events—the settling of a sand grain to the ocean bottom, the tumbling of a rock off a thawing, north-facing mountainside, the death of a small snail—that add up to geologic change.

CLIMBING A MOUNTAIN OF GEOLOGIC TIME

An ambitious hiker who climbs from Nooksack Falls, on the North Fork of the Nooksack River, to the top of Hadley Peak, crowning Chowder Ridge, north of Mount Baker, ascends about 5,000 feet vertically—about a mile straight up—and climbs layered rocks of the Nooksack Formation spanning an age range of about 50 million years. Using this same scale, if our hardy hiker were to ascend the 400 million years of strata widely represented in North Cascades National Park, he would have to climb vertically about 8 miles. And if he were to climb metaphorically through strata representing the time encompassed by all the North Cascade rocks, including some that are about 1,600 million years old, he would have to climb about 32 miles straight up. In contrast, for the youngest episode of North Cascade history, the growth of the Mount Baker and Glacier Peak volcanoes in less than 1 million years, he would only have to scramble up about 100 metaphorical feet. (Figure 4).

THE GEOLOGIC TIME SCALE

Nineteenth-century geologists devised a time scale to represent the relative ages of rocks. The age of the Earth was then much debated, with estimates ranging from a biblical 6,000 years to figures on the order of 40 million years, still far short of the currently accepted age of about 4.6 billion years. Despite disagreements as to the age of the Earth, it was still possible in many places for everyone to agree that one layer of sedimentary rock rested on top of another and that the one on top was younger. As a result, geologists were able to create a relative geologic time scale based on such relationships (Figure 5).

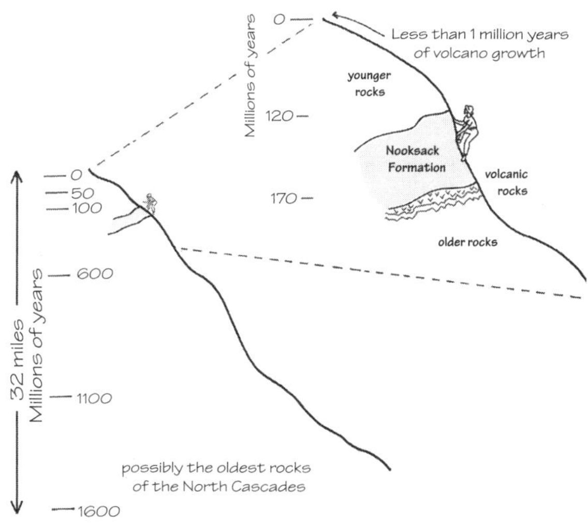

Figure 4. Climber on the mountainside of geologic time (climber not to scale).

No one knew the ages of rocks expressed in years until dating methods using naturally occurring radioactivity were developed in the twentieth century. In this book we avoid using the plethora of relative geologic time terms (all those eras, periods, and epochs illustrated in Figure 5). Instead, we give ages in years whenever possible, while reporting the relative ages in parentheses for those who enjoyed *Jurassic Park* and would like to know a little more about other time periods, such as the Triassic and Cretaceous.

For more information about dating rocks by radioactivity see Geologic Notes 4, 69, and 94.

THE BASIC PATTERN EMERGES

Early explorers in different parts of the North Cascades discovered such a variety of rocks and structures that the overall geologic pattern remained obscure for a long time. Even so, early workers recognized that the Earth's crust in the vicinity of the North Cascades is tilted up to the north (Figure 6). The tilted-up range was, and still is, higher in the north than in the south, and as a consequence, erosion has not only whittled the northern part down more rapidly and bitten deeper into the crust, revealing the older rocks, it has produced greater local relief.

In southern and central Washington, the Cascade Range is made of young volcanic rocks, ranging from 0 to about 40 million years old (Tertiary), erupted from many volcanoes of the Cascade Volcanic Arc (Chapter 6). North of Snoqualmie Pass (Interstate 90), a foundation of older rocks begins to peek out from under this volcanic cover. Farther north, fewer young volcanic rocks are preserved and more of the foundation can be seen. In the area covered by this guide, the few patches of volcanic rocks that remain are mostly preserved in down-dropped fault blocks. All the rest of the volcanic cover has been eroded away, exposing the granitic rocks which once, as magma, invaded the foundation of older rocks and became the reservoirs of molten rock that fueled the volcanoes (Figure 3).

DOMAINS OF THE NORTH CASCADES

The early studies of Peter Misch and his students established a general architecture for the North Cascades' foundation. The range is sliced by two major faults that separate the rocks into three distinct blocks. *Faults* are significant crustal cracks where the rocks on each side have moved relative to each other. The western fault is the Straight Creek Fault; the eastern, the Ross Lake Fault. Because the Ross Lake Fault consists of many fault strands, it is sometimes called the Ross Lake Fault System.

For convenience in this book, we call the three faulted blocks *domains*. From west to east they are the Western Domain, the Metamorphic Core Domain, and the Methow Domain (Figure 9).

The Western Domain consists mostly of sedimentary and volcanic rocks and, although these rocks are complexly faulted and folded, they still retain textures and structures, such as sedimentary bedding, that are typical of their origins on the surface of the Earth. Many of these rocks contain fossils.

The central Metamorphic Core Domain comprises highly squeezed and recrystallized metamorphic rocks which were once at great depths in the Earth's crust. Their

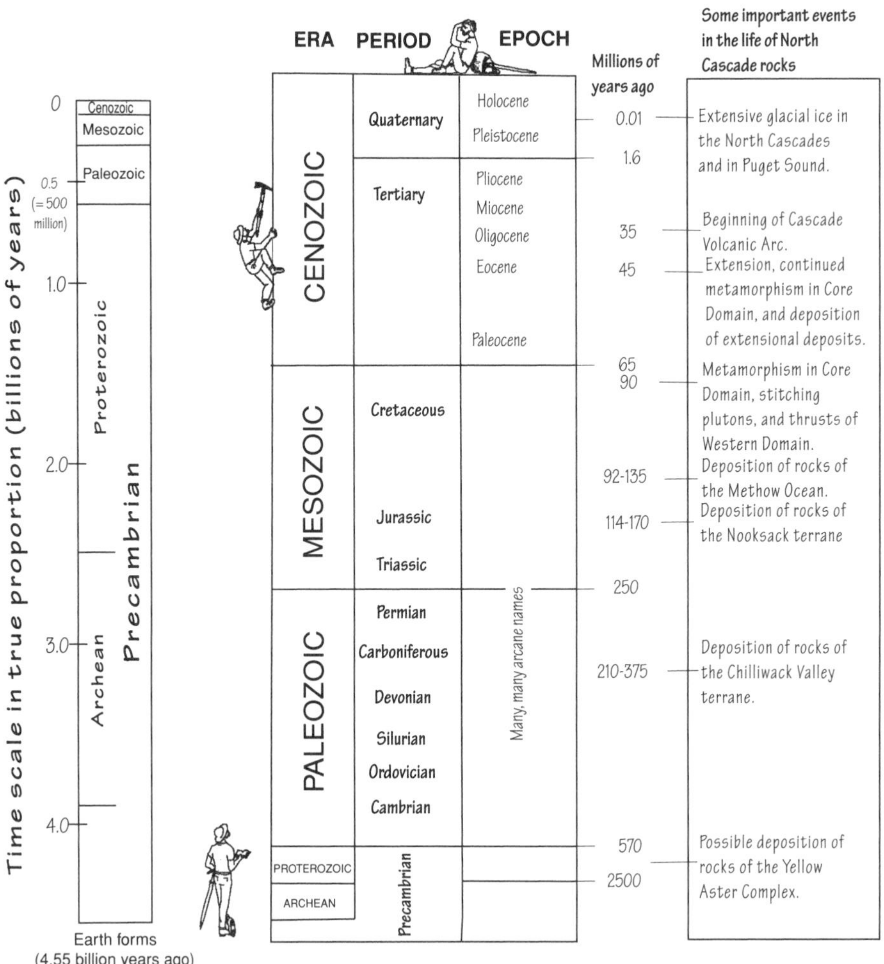

Figure 5. Geologic time scales (much simplified).

story is complex. If they ever had fossils, the fossils have been destroyed by the metamorphism. How these rocks originally formed can only be inferred from their compositions and vague relicts of their original structures and textures.

The Methow Domain is essentially unmetamorphosed sedimentary and volcanic rocks. They are less complex than rocks in the other domains and have fossils, in some places abundantly.

PLATE TECTONICS AND TECTONIC TERRANES

Understanding the basic foundation rocks of the North Cascades became possible only with the birth of the concept of plate tectonics in the late 1960s and early 1970s. Plate tectonics holds that the Earth's outer shell is made up of many large, relatively rigid plates floating on a denser layer of plastic rock, the lower part of the mantle. These plates move about at speeds no greater than a few inches per year. Where plates collide, mountains rise up; where they pull apart, oceans are formed. At the collision zone between major plates in the Pacific Ocean on the one side and the North American continental plate on the other, not only were many of the rocks now seen in the North Cascades born, but they were smashed on a mind-boggling scale as the plates slowly collided or slid by each other along faults.

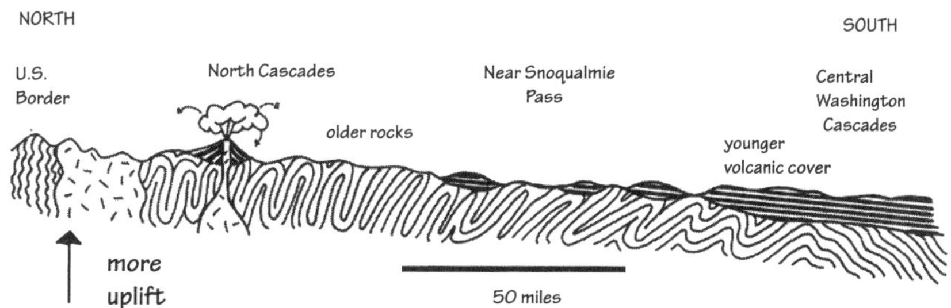

Figure 6. In the North Cascades erosion has stripped off more of the volcanic cover than it has in the southern part of the Cascade Range due to the greater uplift to the north.

An Early Geologist and Early Ideas

Peter Misch (Figure 7), a professor of geology at the University of Washington, made the first real geologic map of part of the North Cascades. Misch was born in Germany, learned geology in the Alps and in the Pyrenees, and had done field work in the northwest Himalayas. When the Nazis took over, he fled to China to teach and study the rocks of Yunnan before coming to the United States at the end of the Second World War. He began his study of the North Cascades in 1948 and, being an avid hiker and mountain climber, reached many remote areas in the North Cascades never before visited by geologists.

Figure 7. Peter Misch

Misch was quick to apply ideas of European alpine geology to the rocks of the North Cascades. He viewed the range as a symmetrical up-folded welt of metamorphosed rocks, thrust out east and west over less deformed and metamorphosed strata (Figure 8). This view has been considerably modified by more recent work, but many of his original observations have proven remarkably correct.

The work of Misch and his many students established a framework upon which all subsequent studies have been built. They were the first to recognize that the North Cascades consisted of three wide belts of rocks (Figure 9), trending north-south, that revealed very different geologic histories. In this book, we refer to these belts as domains.

When Misch and his students assembled the first geologic maps of the North Cascades, they did not imagine that the groups of rocks (the "tectonic terranes" discussed in this chapter) that they had mapped may have traveled great distances before they were assembled where they are today.

Misch constructed a geologic history that assumed that the North Cascades pretty much formed in place, where they are now. Oceans were filled with sediment and volcanic rock that both eventually became metamorphosed. New oceans formed and younger rocks were deposited. Everything more or less happened where the rocks are today. Now we tend to view the geologic scene with the great mobility of the Earth's crust in mind. Geologic terranes may have come from great distances, hundreds to thousands of miles, before they were assembled where they are today. Ironically, Peter Misch was thought to be a radical mobilist in his day because he proposed that some rock units in the North Cascades had been thrust over each other for tens of miles!

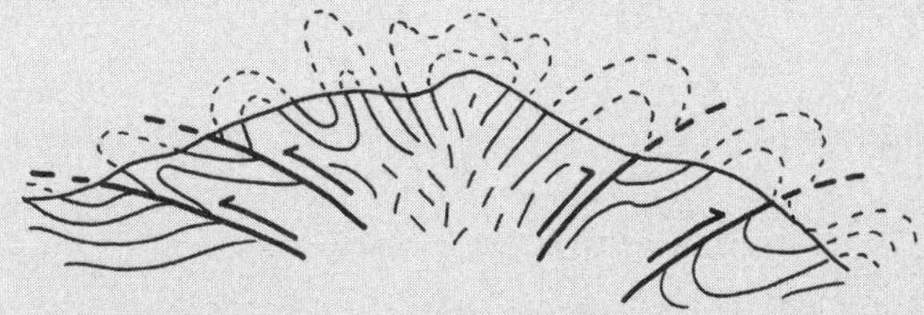

Figure 8. Diagrammatic cross section showing structure of a double-sided, folded and faulted mountain range. Peter Misch thought the North Cascades had this structure, but we now know the picture is more complicated.

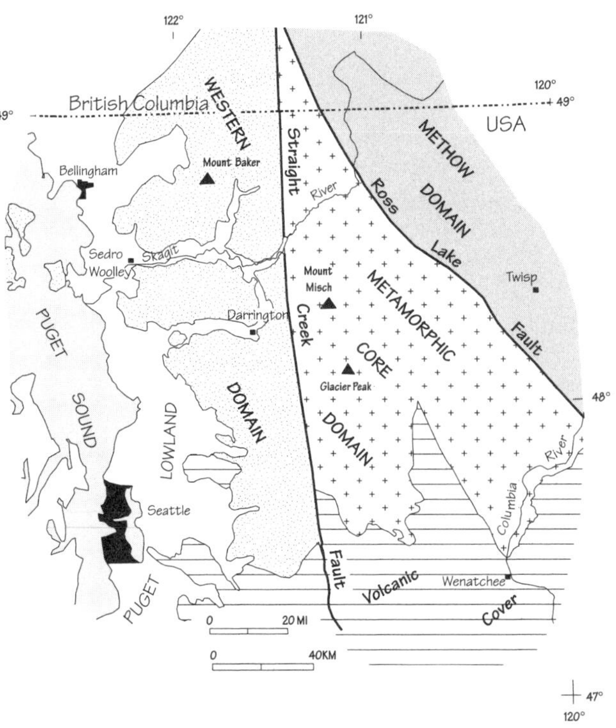

Figure 9. Three major geologic domains—the Western, the Metamorphic Core, and the Methow—make up the North Cascades. Heavy lines are major faults.

Geophysicists have long known that the Earth is composed of three principal layers: a nickel-iron core, an intermediate layer of very dense rock called the *mantle,* and a relatively thin (3 to 50 kilometers) crust of lighter rocks more or less floating on the mantle (Figure 10). Geophysicists discovered this structure by studying sound waves generated by earthquakes as they pass through the Earth. The outermost crust can be further divided into the light rocks of the continents and the heavier, denser rocks of the ocean floors. The distinction between the crust (continental or oceanic) and the underlying mantle is primarily one of chemical composition. Geophysicists also infer another boundary, within the upper part of the mantle, that separates cooler, more rigid rock above from hotter, more plastic rock below. The upper, more rigid rocks—both crust and uppermost mantle—are collectively termed the *lithosphere* and form the plates of the Earth's outer shell.

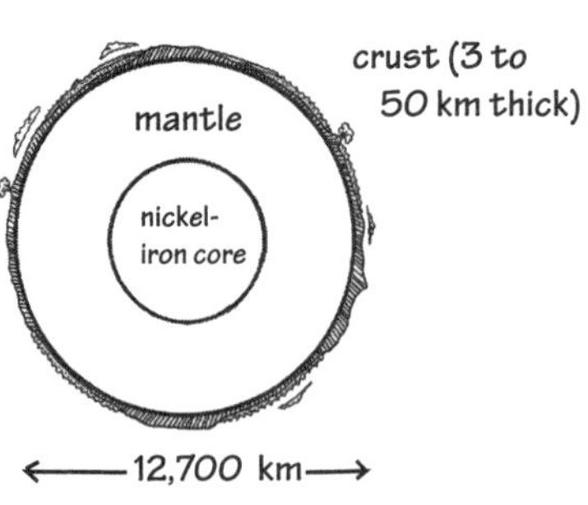

Figure 10. The Earth's crust floats on the denser mantle.

SEA-FLOOR SPREADING

Using their simple tools—compass, hammer, hand lens, and microscope—geologists in the first half of the twentieth century learned much about the continental crust, but little about the oceanic crust. This changed following World War II, when ocean-going geologists adapted sensitive magnetometers developed for anti-submarine warfare for use in seafloor research. This modern technology gave geologists their greatest boost in more than a century of field work. Magnetometer readings revealed that rocks on the ocean floor were magnetized in a startling pattern of symmetrical stripes. The magnetic polarity of every other stripe matched the normal polarity of the Earth today while the alternate stripes had a reversed polarity. This means that if you could isolate a normal stripe from the Earth's magnetic field and hold a compass needle close to it, the north-seeking pole of the needle would align with the

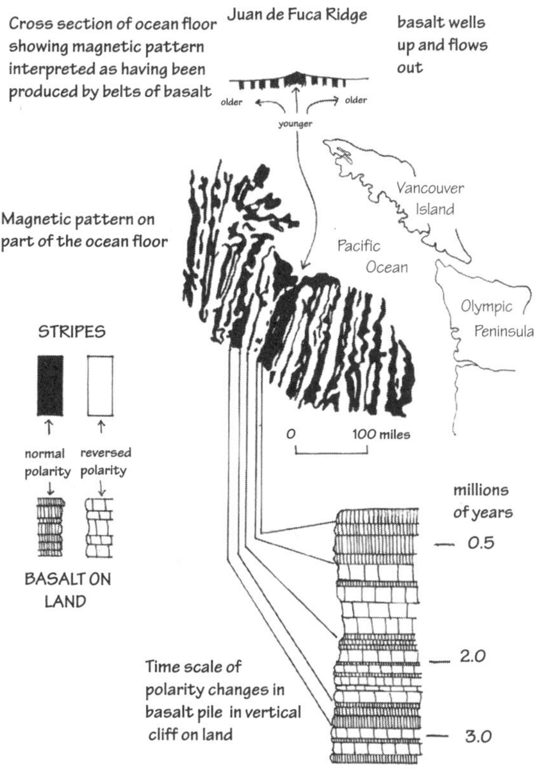

Figure 11. Comparison of magnetic stripes on the seafloor to the magnetic reversals in basalt on the land as seen in a vertical cliff (adapted from Raft and Mason, 1961, and Tabor, 1987).

weak magnet of the rock stripe and point in the same direction as the Earth's north magnetic pole, but if you put the same needle next to a reversed stripe, the north-seeking pole of the needle would point in the direction of today's south magnetic pole. The stripes were of different widths and ran parallel to oceanic ridges. Most remarkably, the patterns of wide and narrow stripes of alternating polarity that occurred on one side of a ridge were mirrored exactly on the other (Figure 11).

The reason for the pattern became known when geologists studying the ocean floor showed their data to geologists who had noticed a similar pattern of alternating polarity in thick piles of surface lava flows. Geologists also had obtained radiometric ages of the lava flows. They knew how many thousands or millions of years ago each lava flow had erupted. The pattern of long and short time intervals represented by the polarity changes in the lavas matched the pattern of the ocean-floor stripes. The scientists concluded that the lavas on land and the ocean-floor basalt had both formed at times when the Earth alternated between normal and reversed magnetic fields. Furthermore, the oldest stripes were farthest from the ridges, suggesting that the floor was growing and spreading by addition of material at the ridges.

As the ocean floor spreads apart, new lithosphere is created. This lithosphere is basaltic crust formed from magma welling up along the widening crack between two crustal plates moving in opposite directions. The polarity of the Earth's magnetic field is recorded in this new crust; hence, the magnetic stripes. The newly formed lithosphere is hot and thus less dense than older adjacent lithosphere; as a result it floats high on the mantle, forming the largest chains of mountains on Earth, the mid-ocean ridge systems. Since ocean-floor spreading continuously creates new lithosphere, and the Earth is not growing, existing lithosphere must somehow be destroyed. The question is where does the existing lithosphere go?

Geophysicists had long observed a concentration of earthquakes along and under volcanic arcs, such as in Japan and the Andes mountains. The sources of the earthquakes, plotted from data gathered over the years, cluster

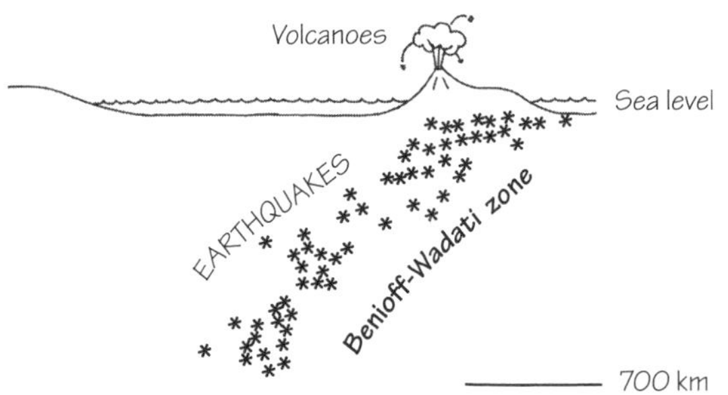

Figure 12. Earthquake locations beneath Japanese volcanic arc (land elevation in this cross section is much exaggerated. After Benioff, 1954).

in planes that dip landward under the mountains (Figure 12). These planes could explain the missing oceanic lithosphere if they were gigantic faults between the disappearing ocean floor and overriding crust. And in fact, the earthquakes in these dipping planes (known as *Benioff-Wadati zones* after their discoverers) are generated as an ocean-floor plate disappears into the mantle under an adjacent plate in a process known as *subduction* (Figure 13). Some of the earthquakes, especially those that are shallow and near the ocean itself, can be very destructive. Many quakes originate within the overriding continental plate, and some very deep quakes appear to originate in the descending plate, where it is breaking apart in the depths of the mantle.

Geologists now know that the Earth's crust is composed of many such *tectonic plates,* floating on the denser mantle. Some are being created and expanded by the injection of upwelling mantle, others are more passive and are moved to and fro by the movements of their neighbors. Much geologic action takes place at the boundaries of tectonic plates. In addition to the locations where the plates spread apart, where new lithosphere is made, and subduction zones, where lithosphere is destroyed, there are boundaries, world-class faults where lithospheric plates slide past one another. Such boundaries are called *transform boundaries.* Perhaps the best-known transform boundary is the San Andreas Fault in California, where the Pacific plate slides north past the North American plate at a speed of about 1½ inches per year.

ROCKS FORM WHERE PLATES COLLIDE

The majority of the rocks in mountain ranges were created where plates collided (Figure 13). Ocean-deposited sediment, turned into sedimentary rock, commonly is scraped off, deformed, and transferred to the overriding

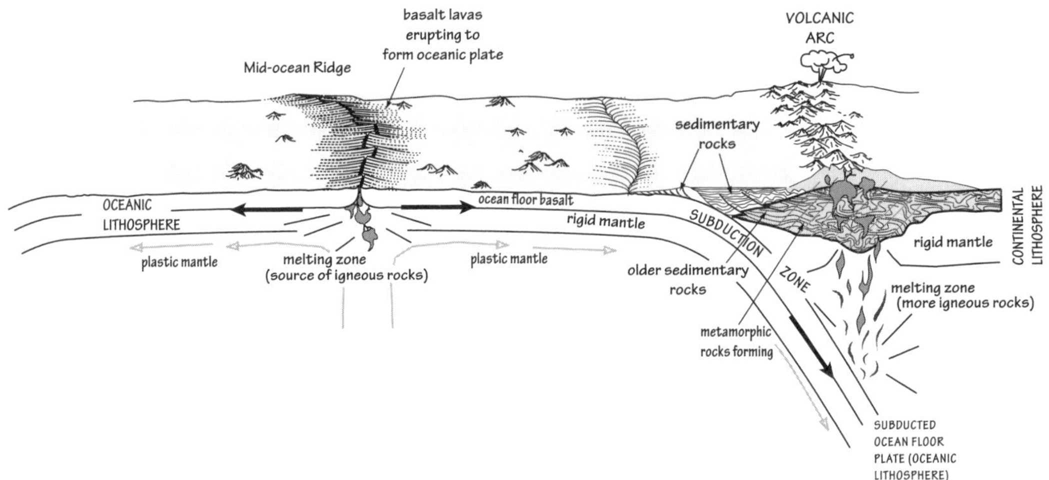

Figure 13. Plate tectonics scheme showing subduction and the formation of the major rock types: igneous, sedimentary, and metamorphic.

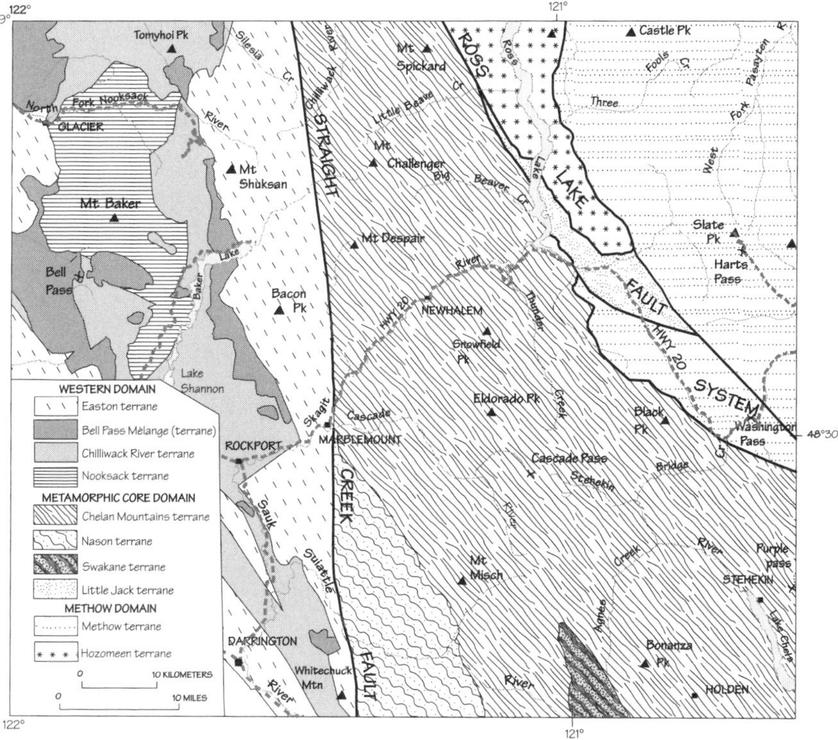

Figure 14. Generalized map of major terranes in the North Cascades. Younger Cascade Arc rocks are not shown. Heavy lines are faults. Heavy dashed lines are roads.

plate. Some ocean-floor sediments are carried down in the subduction zone, where increasing heat and pressure turn them into metamorphic rocks, such as those forming the foundation of the North Cascades. The subducted oceanic plate and overlying sedimentary rocks are rich in water, which is driven off as the rocks move deeper and get hotter. The rising vapors lower the melting point of the already-hot upper mantle and lower crust. As a result, these melt, in the process expanding and becoming less dense than the unmelted rocks around them. The melted rocks begin to rise buoyantly, working their way toward the surface and exploiting any zones of weakness, such as faults. Where the molten rock reaches the surface, volcanoes form. Much igneous rock is created in this subduction process. All around the Earth, chains of volcanoes form arcs that lie above subducting plates, such as the volcanic islands of Japan or the Aleutians.

Off the coast of Washington, the Juan de Fuca plate is slowly subducting under the continent. In Washington, the volcanoes of the Cascade Range (Chapter 6) are located above where the plate has plunged deeply enough to promote melting. In the region that this guide covers, Mount Baker volcano immediately comes to mind. More notorious is Mount St. Helens, which is located in Washington's southern Cascades, where in 1980 magma generated deep in the subduction zone reached the surface to erupt violently and subsequently push up a dome in its crater.

MOVING PLATES AND TECTONIC TERRANES

A consequence of an Earth made up of moving plates is that large slabs of crust can be moved great distances around the globe. Geologists have come to realize that mountain belts can contain rocks that could not have formed in the same setting as their neighbors (see "Making Sense of Metamorphic Rocks and Terranes in General" in Chapter 3). Rocks from the flanks of a volcanic arc on the continent may now be cheek by jowl with rocks of the same age that formed in the middle of a deep ocean with nary a volcano in sight.

Sometimes a piece of one plate breaks off and becomes attached to another plate. Commonly such an orphaned fragment will travel along with its new plate only to shear off on yet another plate. These separated, transported pieces are called *tectonic terranes* or, sometimes, *exotic terranes*. Each of the three domains making up the foundation of the North Cascades is itself a mosaic of several distinct tectonic terranes (Figure 14).

chapter 3

Recognizing the Mountain Mosaic

Division of the North Cascades into the Western, Metamorphic Core, and Methow Domains is a start at understanding North Cascade geology, but as each domain is a mosaic of several terranes, brought together along a variety of faults, the rock-alert traveler needs to know something about the numerous terranes.

Although in this chapter we discuss some of the early events in the history of the terranes, before they were brought together in the North Cascades, we mostly describe the terranes in terms of where and what they are now. The discussions of individual terranes include small iconic maps that are keyed to the larger geologic map (see Plate 2) and help identify their locations.

ROCKS OF THE WESTERN DOMAIN

The Western Domain consists of a folded stack of terranes. From lowest to highest (but not oldest to youngest, as in a sedimentary pile) are:

- The Nooksack terrane: rocks once deposited as bouldery gravel, sand, and mud in a submarine fan flanking a volcanic arc.
- The Chilliwack River terrane: volcanic rocks and sedimentary rocks from a different volcanic arc some 100 million years older than the Nooksack deposits.
- The Bell Pass Mélange: bits and pieces of many things, including deep ocean deposits and pieces of ancient continental crust.
- The Easton terrane: the only well metamorphosed terrane of the Western Domain, derived from deep-ocean sands and muds and underlying basaltic ocean floor.

The southern part of the Western Domain (mostly southwest of Darrington south to Snoqualmie Pass), is underlain by even more terranes, but they are beyond the scope of this guide and are not shown in Figure 14.

To understand the rocks in the Western Domain, we all have to think big. It's not hard to imagine a bedded sequence of sedimentary rocks where each bed on top of the next is younger. Even on the huge scale of the Grand Canyon of the Colorado River our imagination is not too taxed to envision the layers of sediment accumulating over millions of years. The Western Domain in the North Cascades also has a large-scale layering, but each layer is a terrane and the layers were stacked on top of each other by faulting in a relatively short time, geologically speaking—probably on the order of a few million years. Furthermore, the youngest layer is not on top, but at the bottom. The terranes of the Western Domain have been thrust over each other, probably at the western edge of the North American continent as it collided with another tectonic plate (Figure 15). These stacked-up thrust plates have then been folded in a broad arch with the axis trending more or less southeastward under Mount Baker. Erosion has eaten away much of the rock bent into this arch, exposing the lowest rocks in the core, although these are today somewhat obscured by the young volcanic splotch of Mount Baker (Figure 16).

The Western Domain is a stack of odd bedfellows, disparate terranes that formed in different settings and whose previous histories were quite different before they were all put together in what is now the western North Cascades. But the story is a little more complex than this because the geologic forces that stacked these terranes as thrust plates, gave the whole pile at least one more whack, so to speak, and sliced off a chunk of the whole pile and added it to the top. This final slice, the Gold

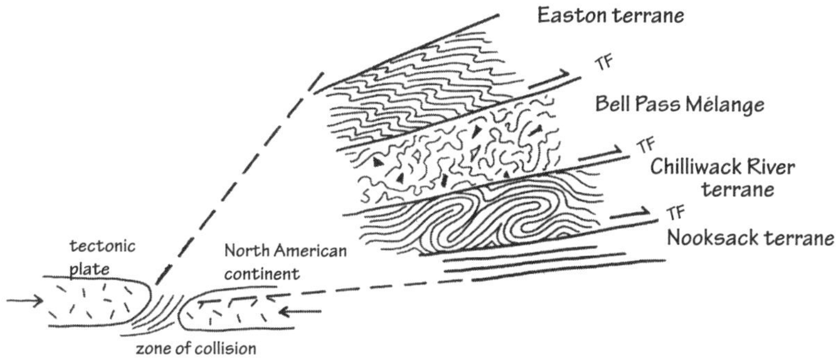

Figure 15. Deformation and emplacement of terranes along thrust faults (TF) during collision of a tectonic plate with the North American plate (as seen in diagrammatic cross section).

Run Pass thrust plate, can be seen at Gold Run Pass, naturally, where rocks of the Chilliwack River terrane are on top of the Easton terrane, not under it, as is the usual arrangement. We all should be spared such a geologic complication, but perhaps that is what is to be expected in a subduction zone where an ocean-floor plate has smashed against a continental plate.

NOOKSACK TERRANE

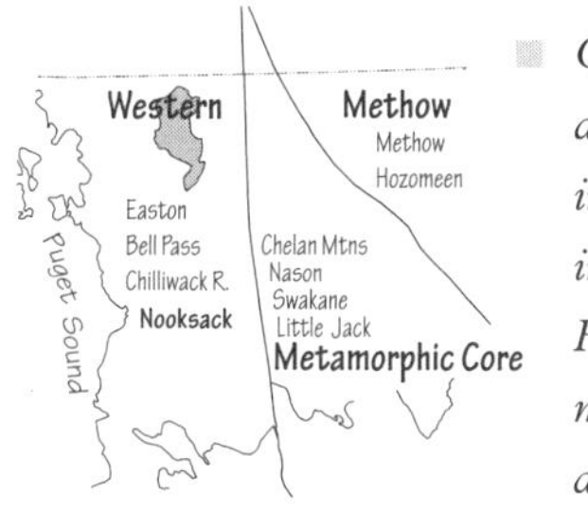

Domains and Terranes

Conglomerate, sandstone, and shale, once deposited in a submarine fan flanking a volcanic island arc. From about 170 to 114 million years old (Jurassic and early Cretaceous). Very little folded and faulted.

The rocks of the Nooksack terrane (also called the Nooksack Formation) are best seen in the vicinity of the North Fork of the Nooksack River. In the Wells Creek area the lower part of the terrane consists of volcanic rocks, probably deposited in the ocean at the edge of island arc volcanoes. These volcanic rocks are overlain by conglomerate, black shale, and sandstone that are many thousands of feet thick. Hikers on the Skyline Divide Trail can view these rocks under their feet (Plate 4A) and spectacularly in the distance to the north across the North Fork of the Nooksack on Excelsior Ridge. Bedding characteristics indicate that some of these sedimentary rocks were deposited in submarine fans washing down off the volcanic island arc.

Many outcrops of the Nooksack sedimentary rocks contain fossils of belemnites, a cephalopod similar to a cuttlefish, with an internal tubelike shell (Figure 17). On Chowder Ridge and elsewhere abundant shells of clams may also be seen. These fossils first indicated to geologists the age of the rocks of the Nooksack terrane. More recently, radiometric analyses (see Geologic Note 4) of individual grains of zircon in the sandstone have confirmed the fossil ages and extended the age range. For

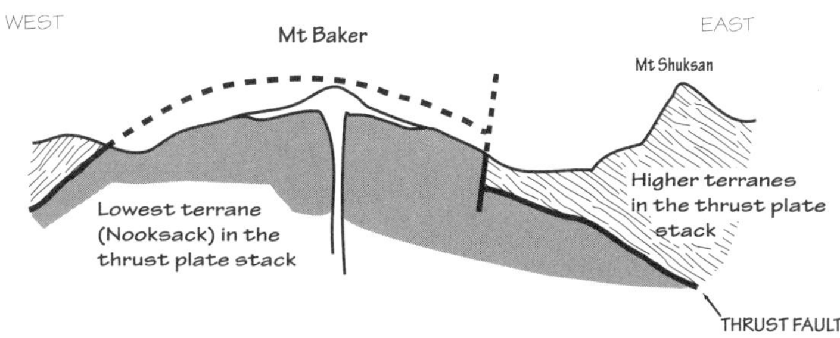

Figure 16. Up-arched stack of overthrust terranes with erosion exposing lowest terrane in core of arch. Dashed line shows location of folded thrust plate before erosion.

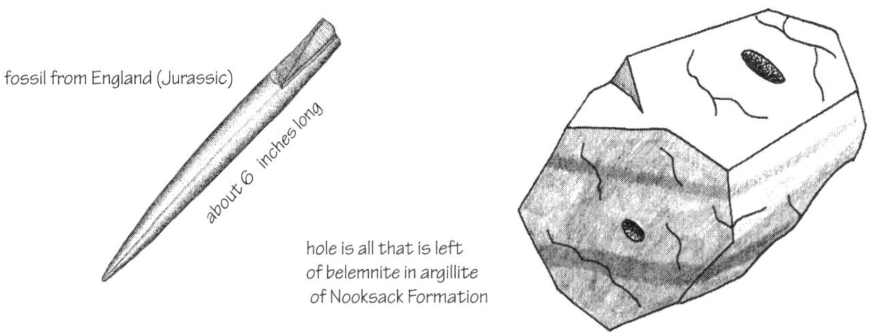

Figure 17. Belemnite fossils

more examples of the Nooksack terrane, see Geologic Notes 74, 84, and 85.

CHILLIWACK RIVER TERRANE

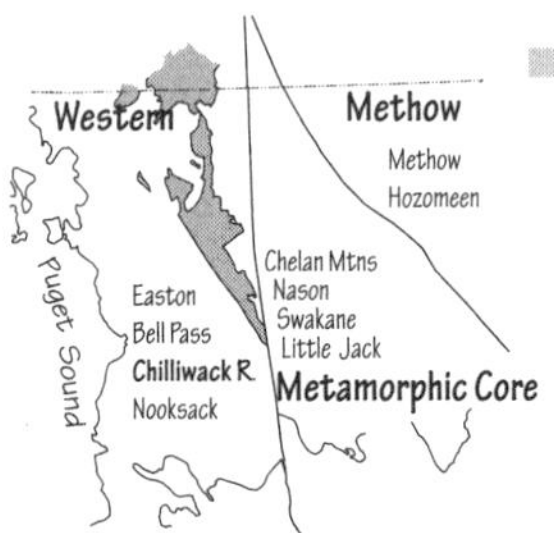

Domains and Terranes

Mildly metamorphosed volcanic and sedimentary rocks from a volcanic island arc or arcs. From 375 to 170 million years old (Devonian to Triassic). Highly folded and commonly upside down.

On top of the Nooksack terrane is the Chilliwack River terrane, named for the Chilliwack River in British Columbia. Volcanic rocks dominate, but sandstone, shale, and scattered limestone are also abundant. We know from many fossils that the Chilliwack River terrane ranges from roughly 375 to 170 million years in age (Devonian, Carboniferous, Permian, and Triassic). As the reader can see from the geologic time scale (Figure 5), the Chilliwack River terrane is older than the underlying Nooksack terrane. Such a topsy-turvy relationship is a sure sign of significant fault movement. Nevertheless, as with the Nooksack, the rocks of the Chilliwack River terrane were also formed as part of a volcanic arc, probably one fringed with limestone reefs.

The Chilliwack River terrane is strongly folded and faulted and somewhat metamorphosed. Many of the shales have changed to phyllites and volcanic rocks have become greenstones. Limestone is metamorphosed to marble, but the recrystallization has not destroyed all the fossils. Careful study of sedimentary-bedding features, such as graded bedding formed in density flows, indicates that broad areas of Chilliwack rocks are entirely upside down. No part of the Nooksack terrane is like that! Chilliwack rocks must have been folded and faulted before, or as, they were shoved on top of Nooksack rocks, which, for the most part, are only gently folded and almost always right-side up (Figure 18).

For more information on the Chilliwack River terrane, see Geologic Notes 1, 76, 79, and 86.

BELL PASS MÉLANGE

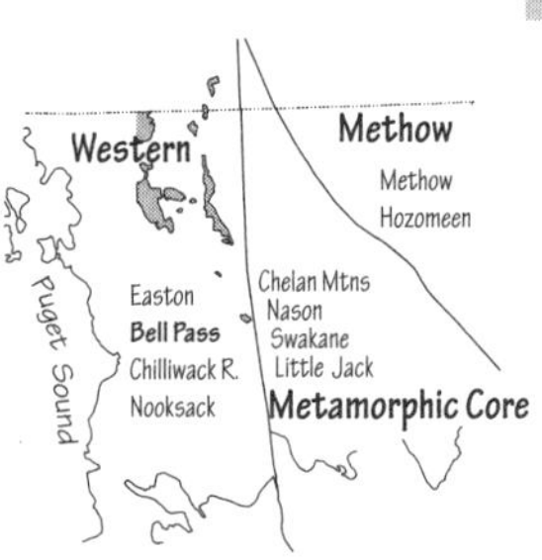

Domains and Terranes

Bits and pieces of many things, including deep ocean deposits, pieces of ancient continental crust, and pieces of the mantle. Ages of constituents range from probably more than 570 (Precambrian) to about 200 (Triassic) million years old.

On top of the Chilliwack River terrane rests the Bell Pass Mélange. *Mélange* is French for mix, and geologic mélanges are very mixed rocks; they are hodge-podges on a grand scale. Much of the Bell Pass Mélange is made up of sedimentary rocks, including sandstone, shale, and chert. Basalt is also abundant in some areas. The chert, shale, and basalt are not only mixed, but highly broken up. The continuous layers of bedding are gone; only blocks and pieces remain (Plate 5C). The rocks look faulted, or, more expressively, smeared. Blocks of hard basalt or chert stick out here and there as resistant knobs. But the mélange contains exotic rocks as well, things we do not expect to find mixed into unmetamorphosed

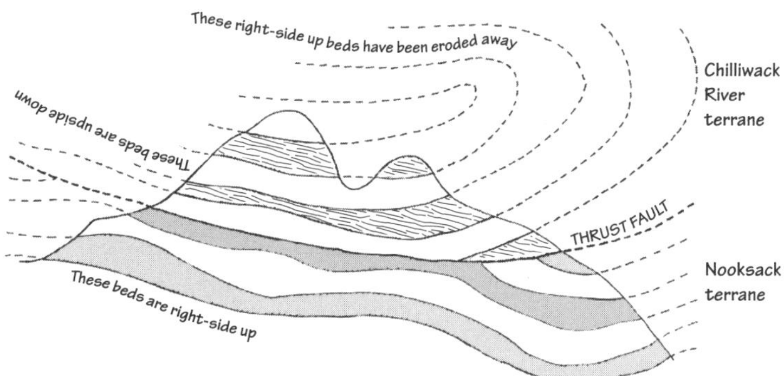

Figure 18. Before the Chilliwack River terrane was thrust over the Nooksack terrane, its beds were folded upside down.

sedimentary or volcanic rocks. Exotic blocks in the mélange are metamorphic rocks, such as gneiss and schist, which came from deep in the Earth's crust, and ultramafic rocks—dark colored, rich in iron and magnesium—which came from deeper in the mantle. The sedimentary parts of the Bell Pass Mélange are mostly oceanic in origin, but many of the exotic blocks are not (see "Making Sense of Metamorphic Rocks and Terranes in General" in the following section).

The blocks or "knockers" of gneiss (being hard, they resist the knocking of a geologist's hammer) in the Bell Pass Mélange were called the Yellow Aster Complex by Peter Misch, who named the formation for Yellow Aster Meadows, which is underlain by a large slab of gneiss (Plate 3A). The gneiss was formed deep in the crust, at relatively high temperatures. Associated with it are igneous rocks. Both the gneiss and igneous rocks have been smashed and further metamorphosed at cooler temperatures. The origin of much Yellow Aster gneiss is obscure, but on Park Butte, gneiss rich in calcium minerals and associated with marble indicates that some Yellow Aster gneisses were sedimentary rocks before high-temperature metamorphism. The association of metamorphosed sedimentary rocks and plutonic igneous rocks is suggestive of a continental setting, although it is surprising to find such rocks amidst the relatively unmetamorphosed oceanic rocks that comprise most of the Bell Pass Mélange.

Even more startling in the mélange are blocks of ultramafic rock. They began existence in the mantle (see Chapter 2), deep in the Earth. Most ultramafic chunks in the Bell Pass Mélange are small, a few feet to a few hundreds of feet across. The largest block of ultramafic rock in the Bell Pass Mélange is the Twin Sisters Mountain massif, made mostly of *dunite,* a rock consisting of the mineral olivine. Except for the remarkably unchanged Twin Sisters dunite, many of the ultramafic blocks are now partly or entirely made of serpentinite.

(For more examples of the Bell Pass Mélange and its disparate parts see Geologic Notes 54, 57, 58, 59, 79, and 90.)

Sand in the Sea: How Does It Move?

Geologists had long been puzzled by the existence of sandstone beds that seemed to have been deposited in fairly deep ocean. How did the sand get so far out into the quiet water of the deep ocean? Experiments, direct observation of the ocean floor, and theoretical calculations show that the sand is carried in a dense slurry of turbid water that can flow like a stream on the ocean floor. These density flows form where some near-shore event, such as a submarine landslide, a storm, or an earthquake, stirs up sediment and water, which then rapidly descend the sloping ocean floor. As the density flow slows down, the coarser particles settle out first, so that the final sandstone bed is graded from coarsest at the bottom to finest at the top. When the density flow comes to rest, its sandy bed presses down into underlying shales unevenly, producing a lumpy bottom. Many sandstone beds in the Nooksack Formation are graded; many have lumpy bottoms. Geologists can use these features to tell which side of the bed was originally up. Very useful, indeed, where rocks are highly folded.

EASTON TERRANE

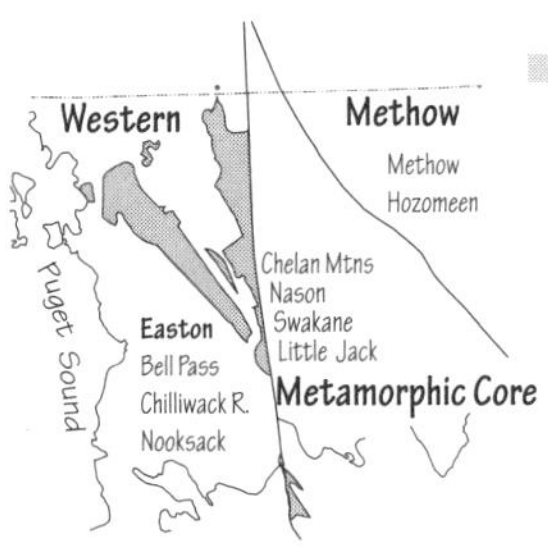

Domains and Terranes

Formed from deep-ocean sand and mud and underlying ocean-floor basalt about 150 million years ago (Jurassic). Metamorphosed at especially high pressure to the Darrington Phyllite and Shuksan Greenschist respectively.

The uppermost terrane in the Western Domain, the Easton terrane, was originally oceanic basalt and overlying deep-ocean mud and sand. The basalt became what has long been called the Shuksan Greenschist, and the overlying sediments became what is known as the Darrington Phyllite. The metamorphosed basalt is not everyday greenschist. Shuksan Greenschist locally contains some unusual blue amphiboles, and the phyllite contains the uncommon mineral lawsonite. We know from experiments and geologic relations elsewhere in the world that rocks with these minerals (called *blueschists* if the blue amphibole is abundant) form only where rocks are buried deeply (to mantle depths, in fact) in a relatively cool environment and then regurgitated relatively rapidly. These conditions are most easily met where plates collide in a subduction zone. If the rocks had remained buried for a long time (tens of millions of years) they would have become hot enough for more ordinary green amphiboles and other minerals of high-temperature metamorphic rocks to have formed. Blueschists are a sort of geologic Baked Alaska.

Making Sense of Metamorphic Rocks and Terranes in General

A lot of things can happen to rocks after they have formed—moving about on tectonic plates, folding, and faulting—and when geologists are confronted with rocks that have also been heated, squeezed, and recrystallized (metamorphosed), they might well wonder what sense can be made out of them.

An alternative to getting bogged down in the details is to look at the big picture beyond the scrim of geologic events. Most rocks are born in one of three major settings (Figure 19), and even after considerable terrane shuffle and metamorphism, these birthplaces can be recognized in rocks of the North Cascades:

1 *Oceanic rocks* are born in the deep ocean, mostly well out to sea, far from the debris shed by continents and volcanic arcs. The sediments are muds, silts, fine-grained sands, and siliceous (quartz) oozes. Many are made up of the shells of tiny marine plankton, radiolarians, whose siliceous skeletons accumulate slowly on the deep ocean floor. They become radiolarian cherts (Geologic Notes 30, 103). Associated with these deposits are basalt erupted from oceanic ridges and oceanic islands like Hawaii. Also, not uncommonly associated with the oceanic rocks are pieces of the underlying mantle, that is, ultramafic rock. After metamorphism, many oceanic sedimentary rocks become mica schists; the chert becomes distinctive mica-quartz schist or quartzite; and oceanic basalt becomes dark amphibolite (hornblende-plagioclase rock). Ultramafic rock may recrystallize to a variety of somewhat esoteric schists, but commonly becomes slippery green serpentinite.

2 *Arc rocks* are rocks derived from volcanic arcs. Arc rocks are both volcanic and sedimentary, because not only do volcanoes grow as the arc develops, but they erode, and the eroded debris is deposited along the arc as sediment fans. If the arc is in the ocean or adjacent to it, the erupted and eroded debris spreads out under the sea as submarine fans. These fans include layers of volcanic rocks alternating with sandstones and shales made up of volcanic-rock grains. Other rocks that may be present are limestone from reefs that may form around volcanic islands, and crystallized magma, mostly granitic igneous rocks, from the chambers that fed the arc volcanoes. Metamorphosed arc rocks have very little mica-quartz schist (at least from near the flanks of the volcanoes, where siliceous oozes have no time to accumulate) and have a greater variety of minerals than metamorphosed oceanic rocks, reflecting a greater variety of volcanic rocks. The limestone reefs metamorphose to marble. In the North Cascades, some metamorphosed arc rocks include metamorphosed conglomerate (or *metaconglomerate*), a certain clue that the volcanoes were eroded by streams. Magma chambers beneath the arc volcanoes crystallize

Ned Brown (Professor of Geology at Western Washington University and an ardent fan of the Shuksan Greenschist) and his colleagues have determined that the original ocean-floor basalt and overlying mud and sand may have formed about 160 million years ago (Jurassic). They were metamorphosed some 40 million years later, in the Cretaceous Period.

For more on the Easton terrane, see Geologic Notes 18, 48, 80, and 91.

ROCKS OF THE METAMORPHIC CORE DOMAIN

Sorting out the original materials that were once stewed deep in the crust and are now exposed in the Metamorphic Core Domain is challenging but not without rewards. First, sands and muds that have been well recrystallized are hard and resistant to erosion (Chapter 7). They stand up in some of the world's most awesome alpine scenery. Many of the North Cascades' highest, most rugged, and challenging peaks are carved from metamorphic rocks. The geologist, seeking the ultimate origin and history of these metamorphic rocks, gets to savor some wonderful scenery. And figuring out how each rock began and what it went through is fun for those who like puzzles and unsolved mysteries.

Fortunately for puzzled geologists, the birthplaces—oceanic, arc, and continental (see "Making Sense of Metamorphic Rocks and Terranes in General")—of the major rocks (Figure 19) in this domain are generally recognizable even after thorough metamorphism to schist and gneiss at great depths and pressures. The four terranes of the Metamorphic Core Domain all contain representative rocks from the major birthplaces.

to become arc-root plutons. These granitic rocks may be squeezed and recrystallized by metamorphism, but their origins are still recognizable.

3 *Continental rocks* are rocks eroded from old continental landmasses. This group is a little more tricky to recognize. If the continents are highly eroded, then the debris carried into the ocean, to form sandstones, shales, and conglomerates, is rich in fragments of metamorphic and granitic igneous rocks. Naturally some rocks will be derived from arcs and continents and even uplifted oceanic rocks, and such mixtures are not so easily classified. After metamorphism, continental rocks may only be distinguished from arc rocks, if at all, by subtle differences in chemical composition.

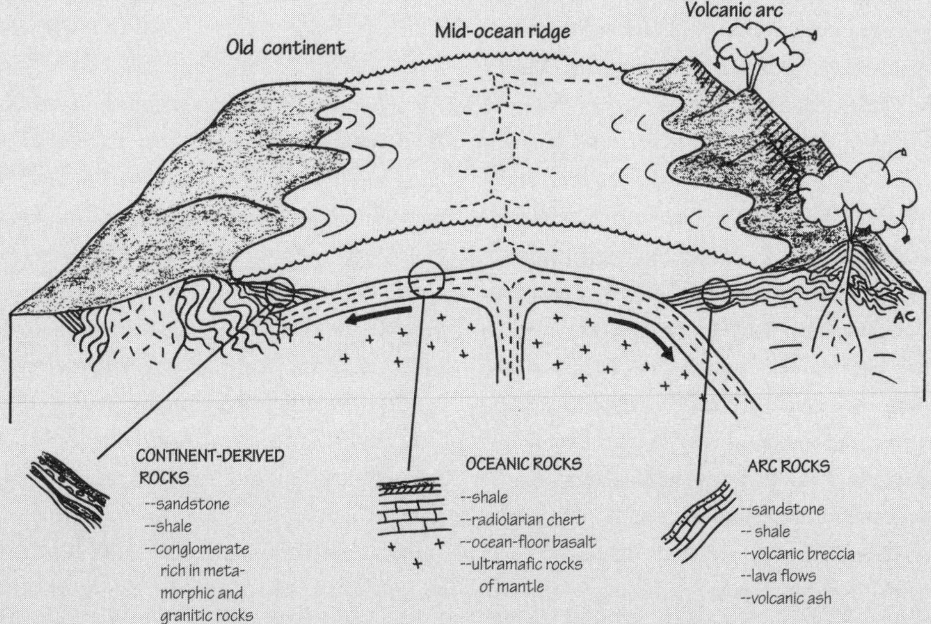

Figure 19. Rocks are categorized as continental, oceanic, or arc depending on their place of birth.

CHELAN MOUNTAINS TERRANE

Domains and Terranes

Metamorphosed deep-ocean deposits and metamorphosed volcanic arc rocks and sediments. Includes metamorphosed arc-root plutons and overlying arc rocks that are 220 million years old (Triassic).

The Chelan Mountains terrane contains both metamorphosed oceanic and arc rocks. The ocean-born rocks characteristically contain metamorphosed basalts; they are now amphibolite (hornblende and plagioclase rock) or hornblende schist. Some are metamorphosed cherts—called *metachert, mica-quartz schist,* or *metaquartzite.* Metamorphosed sandstones and shales are also present, now as mica schists. In addition, the ocean-born rocks contain marble derived from marine animals, but no fossils are preserved. Scattered small pods of metamorphosed ultramafic rocks indicate that bits of mantle were incorporated into the ocean-floor rocks at some stage in the development of the Chelan Mountains terrane, such as when the oceanic plate riding on the mantle swept into a subduction zone prior to metamorphism or when pieces of uplifted mantle, crunched up at the edge of an earlier subduction zone, slid into the ocean deeps. The ocean-born metamorphic rocks have various local names such as the Napeequa Schist and Twisp Valley Schist.

In the Chelan Mountains terrane, rocks born of a volcanic arc are mostly metamorphosed sandstone and conglomerate. The sandstones are now mica schists. The metamorphosed conglomerates—metaconglomerate—look like schists, but many still retain their characteristic cobbly look on some surfaces. The metamorphosed arc rocks also include hornblende schist and amphibolites from basaltic volcanic rocks, along with white mica schists derived from explosively erupted volcanic ash. These rocks are known all together as the Cascade River Schist.

Especially impressive in the Chelan Mountains terrane are remnants of the plutonic root of the old volcanic arc, known to geologists as the Marblemount pluton or plutons, which form high ridges in a long southeast-trending belt (see "MM" and "MD" on Plate 2). This rock forms the imposing west ridge of Lookout Mountain east of Marblemount, makes up Le Conte Mountain and Old Guard Peak on the South Fork of the Cascade River, and holds up imposing peaks to the southeast. We know from radiometric dating that the volcanic arc fed by the magma that formed the Marblemount plutons thrived about 220 million years ago.

Metaconglomerate in the Cascade River Schist contains boulders of the Marblemount plutons, indicating that streams were eroding the cooled, solid pluton when the arc rocks were deposited.

Many rocks of the Chelan Mountains terrane have been so thoroughly metamorphosed and laced with younger granitic igneous material, some as young as 45 million years old (Eocene), that we include them in the Skagit Gneiss Complex, described in Chapter 4.

For more information on the ocean-born rocks of the Chelan Mountains terrane, see Geologic Notes 19, 22, 67, and 137. For more information on metamorphosed volcanic arc rocks, see Geologic Notes 61, 65, 68, and 71.

NASON TERRANE

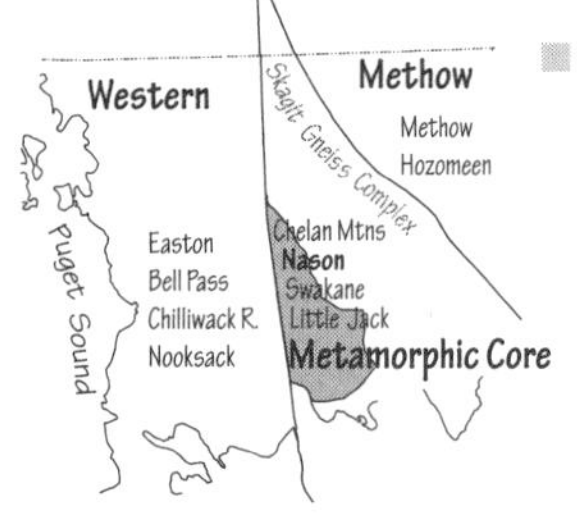

Domains and Terranes

Mostly metamorphosed sandstone and shale derived from a distant volcanic island arc. Original sediments deposited about 120 million years ago (Cretaceous).

The Nason terrane underlies much of the Glacier Peak Wilderness and parts of British Columbia. It does not crop out in the North Cascades National Park Complex (see Plate 2). The Nason terrane appears to be metamorphosed submarine fans derived from distant arc volcanoes. The rocks are now mostly mica schist, hornblende-mica schist, and amphibolite, and have been called the Chiwaukum Schist after exposures in the Chiwaukum Mountains, southeast of Stevens Pass (south of the area covered by this guide). Large tracts of the Nason terrane are gneisses derived from the Chiwaukum Schist and metamorphosed granitic materials. Recent analyses of individual sedimentary zircon grains in less metamorphosed rocks correlated with the Chiwaukum Schist by Professors Edwin H. Brown and his colleague, George E. Gehrels, indicate the fans were deposited about 120 million years ago

(Cretaceous). The fragments of Nason terrane in British Columbia appear to have gotten there by faulting along the Straight Creek Fault (Chapter 5).

For more information on the Nason terrane, see Geologic Note 108.

SWAKANE TERRANE

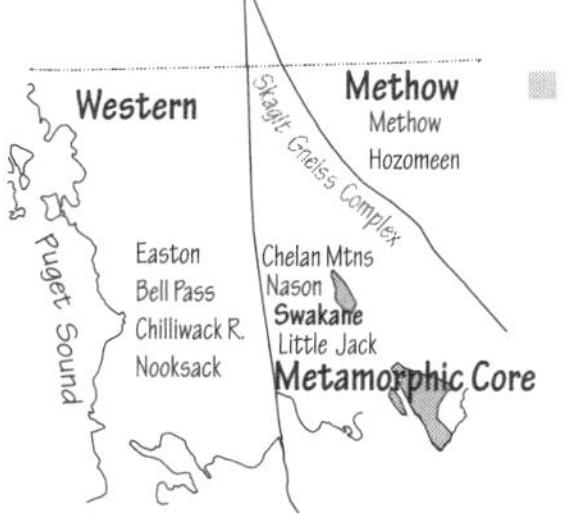

Domains and Terranes

Metamorphosed sandstones containing very old (Precambrian) and relatively young (Late Cretaceous) zircons that suggest an unusual geologic history.

The Swakane terrane, made up of the Swakane Biotite Gneiss, has been a significant geologic mystery. The best place to see the gneiss is along the Columbia River between Wenatchee and Chelan, just south of the area described in this book. Only a small part of the Swakane shows up on the geologic map (see Plate 2; Figure 20) in this book.

The mineralogy and chemical composition of most Swakane Biotite Gneiss suggest that it could have begun either as a sandstone or as a volcanic rock, but not likely both, because it is monotonously uniform overall. Interlayered volcanic and sedimentary rocks, even when thoroughly metamorphosed, are more varied in look and composition. Many years ago, pioneering geochronologist Professor James Mattinson and colleague Professor Clifford Hopson separated large numbers of zircon crystals from samples of the Swakane Gneiss. Analyses of radioactive uranium and lead isotopes (Geologic Note 4) in these crystals indicated that the zircons are very old, perhaps as old as 1,600 million years (older Precambrian). If the original rock was volcanic, then the crystals formed when the volcanoes were erupting.

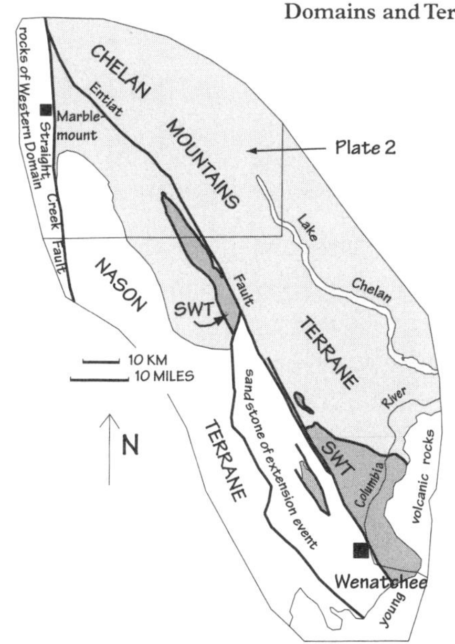

Figure 20. Location of the Swakane terrane (SWT), a mystery terrane of the North Cascades.

But if the rock was a sandstone, then the zircons were most likely sand grains, older than the rock which contains them. The zircon crystals may have been eroded from an older continent. Using new methods to analyze single zircon crystals in the rocks, Dr.Jennifer Matzel and colleagues have determined that the Swakane Gneiss is derived from sandstone that contains zircons as young as 73 million years (Late Cretaceous), although many of the zircons are very old. Cretaceous sandstone, derived in part from Precambrian terrane, was buried deep and metamorphosed, then uplifted and exposed to the surface when the rocks over them slid off about 50 million years ago (Eocene). This abreviated life history for the Swakane does not fit well into other geologic events in the North Cascades as described in this book. Some mystery remains in the Swakane story, and it illustrates well the uncertainty in our geologic speculations. See Geologic Note 116.

LITTLE JACK TERRANE

Domains and Terranes

Metamorphosed sediments deposited at the toe of a submarine fan derived from a volcanic arc. Contains blocks of metamorphosed mantle.

Another mystery terrane is the Little Jack terrane, which is mostly exposed in the Ross Lake Recreation Area. Fine-grained mica schist and layers of fine-grained amphibolite make up the Little Jack terrane. Most of the rock probably began as deep-sea sediments at the toe of a submarine fan. In these metamorphosed sediments are blocks of ultramafic rock; that is, metamorphosed mantle. This peculiar assemblage suggests that as the sediments accumulated, blocks of mantle material, perhaps previously uplifted at a subduction zone, slid off into the ocean.

The Little Jack terrane might be metamorphosed rocks of the Methow Domain, except that the Methow rocks contain no ultramafic blocks.

ROCKS OF THE METHOW DOMAIN

The east-bound traveler, leaving the canyons that drain the hard rocks of the Metamorphic Core Domain, and looking out across the broad valley of the Methow, is struck by the change in scenery: from alpine to pastoral, from cascading water to broad meandering river, and from rocky peaks sheltering perpetual snowfields and glaciers to rocky but rounded and dry hillsides. Climate is the main actor in this scenic transformation. By the time previously water-rich clouds arrive over the valleys of the Methow River area, they have dropped much of their moisture as rain and snow in the process of rising over the North Cascades (Figure 2).

But bedrock also plays an important supporting role. Much of the Methow Domain consists of unmetamorphosed marine sandstone and shales deposited in submarine fans at the edge of a continent. Overlying these continent-derived rocks are sandstone and conglomerate deposited by streams and mixed with volcanic rocks of an arc. The unmetamorphosed sedimentary and minor volcanic arc rocks surrender to erosion more readily than their harder, metamorphic cousins to the west (Chapter 7). Included in the domain is the ocean floor, of slightly metamorphosed basalt, which received these sediments. Compared to the jumbled and broken beds of rocks in the domains to the west, the beds in rocks of the Methow Domain display a great deal more order and predictable behavior. Over broad areas, sedimentary beds are stacked, one on top of the other, in the same order they were deposited (Plate 3B). Well, not totally, because fossils and sedimentary structures indicate the rocks have been strongly folded, and in some areas faulted (see "Breaking and Folding the Rocks of the Methow Domain"). Nonetheless, the sequence in general represents deposits from an ocean basin that gradually filled up. The rocks of the Methow Domain were lovingly studied by Julian Barksdale, Professor of Geology at the University of Washington, from 1936 until his death in 1983. We have since learned a few more details, but the basic framework was established by Barksdale.

HOZOMEEN TERRANE

Domains and Terranes

Ocean-floor basalt, deep-ocean sandstone and shale, with lots of chert. Formed between 350 and 220 million years ago (Mississippian to Triassic).

The Methow Domain began as ocean floor formed between 350 and 220 million years ago. The ocean floor was basalt overlain by some deep-ocean sandstone and mud, along with lots of chert. Some of the basalt erupted at a mid-ocean ridge, some erupted as oceanic islands like Hawaii, and some may have erupted as part of a volcanic island arc. We call this ocean floor the Hozomeen terrane. The rocks of the Hozomeen terrane do not fit well our bucolic description of the Methow scenery—far from it. They stand out ruggedly in the high mountain ridge of Crater and Jack Mountains and peaks to the north.

For more information about the Hozomeen terrane, see Geologic Notes 12, 14, 40, and 41.

METHOW TERRANE

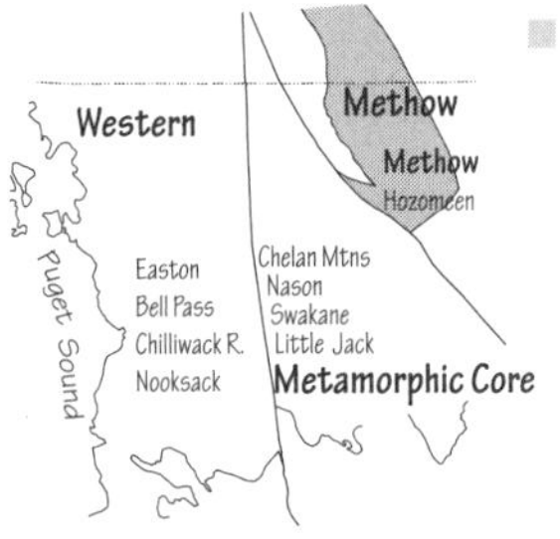

Domains and Terranes

Sandstone and shale from ocean sediments that filled the ancient Methow Ocean about 200 to 100 million years ago (Jurassic and Cretaceous). Sandstone and conglomerate deposited by rivers and streams on top of Methow Ocean sedimentary rocks, and volcanic rocks from a short-lived volcanic arc about 100 million years ago (Cretaceous).

For over 100 million years after the Hozomeen terrane formed, it remained an ocean floor, on which sand and mud collected. In Canada, rocks record the growth of a volcanic arc on this old Hozomeen ocean floor about 180 million years ago (Jurassic), but rocks in the area of this guide only record the continued filling of the Methow

Ocean with mud, sand, a few density flows of pebbles and cobbles, and perhaps a few layers of volcanic ash blown down from the volcanic arc in Canada. By about 110 million years ago (Cretaceous), the Methow Ocean had pretty well filled up. The youngest deposits of this period are shallow-marine sandstones. The Hozomeen terrane was buried.

Tectonic plate movements began (or continued),

Breaking and Folding the Rocks of the Methow Domain

Almost all sedimentary rocks begin life as flat layers of rock deposited on the flats of a river bottom, a beach, or the ocean floor. The very old layers of rock revealed by the Grand Canyon are still lying nearly flat. Yet the traveler in the Methow Domain will notice that most of the layers of sedimentary rock are steeply tilted. Some are even vertical (Plate 3B). Like the rocks of the Western and Metamorphic Core Domains, the Methow rocks have been bent by plate collision, but unlike those other rocks they were never carried down beneath another plate by a subducting slab or completely smashed between two colliding plates. As a result, their deformation is minor by comparison, a sort of fender-bender near the main crash.

Rocks somewhat back from the major plate collision, or adjacent to a great fault where the plates rub past each other, are commonly somewhat crumpled and compressed. If they have well-developed layering, as most sedimentary rocks do, they typically deform in a predictable pattern as upper layers slide over lower layers along planes of weakness such as shale beds. Note that a geologist would have a difficult time recognizing such a fault because the stratigraphic order of the beds would not be changed. But, in places, such faults, which mostly parallel the layers, break up across the bedding into the overlying, harder, sandstone layers, going from the lower, weaker shale beds across the harder beds to an upper bed (Figure 21). Where the faults break across the stiffer beds, folds form as older beds are shoved over younger beds. Such folds are called fault-bend folds. The size of the fold corresponds to the thickness of the stiff layer or layers separating the weak, sliding, layers.

Most of the folds in the Methow Domain appear to be fault-bend folds. As the Hozomeen terrane was thrust eastward over the rocks of the Methow Ocean, at the west edge of the Methow Domain, subsidiary thrust faults parallel to the layers formed in the ocean sedimentary beds as well. In places they stepped up from lower weak beds to higher weak beds and thus stacked the rocks of the Methow Ocean and the Pasayten Group on top of themselves. The step-ups formed across very thick stronger beds and the corresponding fault-bend folds are quite large as well. In some cases the strong layer included almost all of the rocks of the Methow Ocean—a mile or two thick! The resulting folds measure miles and miles across.

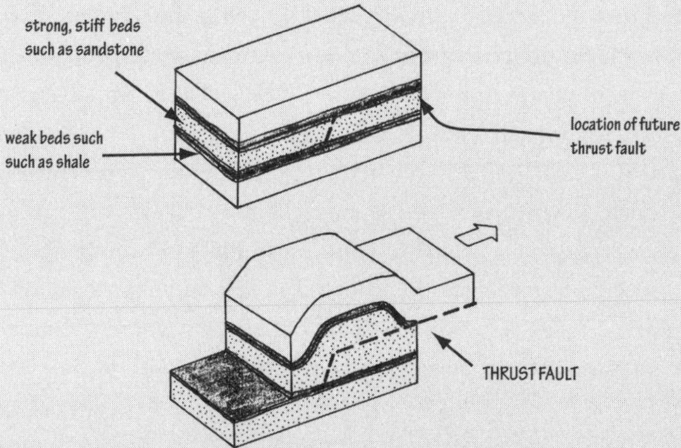

Figure 21. Formation of a fault-bend fold in sedimentary rocks of the Methow Domain. The scale of such folds varies. Fault-bend folds may be visible in an outcrop, on a mountainside, or even across several mountains.

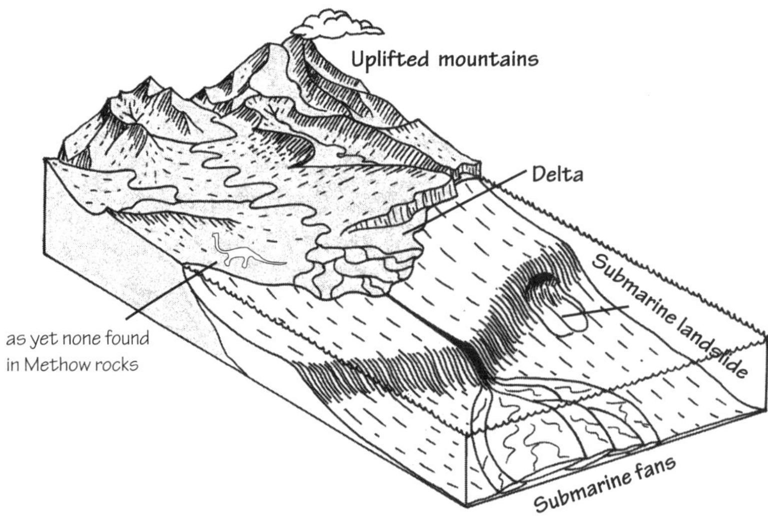

Figure 22. Setting for the accumulation of sediments in the Methow Ocean (after a drawing by David G. Howell)

pushing up the ocean floor west of what is now the Methow area. Some of the Hozomeen terrane was faulted up and shoved eastward, overriding the deposits of the Methow Ocean. About 100 million years ago, the ocean deepened, probably because its western edge was depressed by the weight of the thrust sheet of the Hozomeen terrane, much as the flexing of a diving board bent by the weight of a diver standing on its end. As the upthrust Hozomeen terrane rose above the ocean, erosion tore it down. The now-deep ocean to the east filled with sand, mud, and pebbles flushed off the uplifted and now eroding thrust sheet. This debris is particularly rich in pebbles and sand-sized grains of chert, which contrast with the quartz, feldspar, and mica grains of sand eroded from uplifted granitic batholiths on the east side of the now rapidly filling ocean. All these sediments became the "Rocks of the Methow Ocean" (see Plate 2), which are best seen near Harts Pass and, farther north, in the Pasayten Wilderness. Much of the sediment was deposited by density flows (see "Sand in the Sea: How Does It Move?"). Fossilized seashells help establish the history of the Methow Ocean by indicating the age of the strata and the depth of water in which these organisms lived.

By about 95 million years ago (still in the Cretaceous), the Methow Ocean had all but filled up. Rivers and streams from adjoining highlands poured out mud, sand, and gravels onto the drying ocean deposits, capping the ocean basin with nonmarine materials. In addition, volcanic rocks were deposited on the flanks of a new volcanic arc rising along the east side of the filled ocean. The composition and even the appearance was probably much like the Cascade Volcanic Arc today (Chapter 6). We call all these younger sedimentary and volcanic rocks the Pasayten Group (see Plate 2). A few small granitic plutons—the roots of the volcanoes—intrude the Pasayten Group and Rocks of the Methow Ocean. Some of these younger materials were probably deposited on one or more deltas at the edge of the ancient ocean (Figure 22), a scene not unlike that of the lower Skagit or Fraser Rivers today. The Pasayten Group crops out along Highway 20 north of Winthrop and in the cores of big downfolds in the Pasayten Wilderness.

For more information about rocks formed in and after the Methow Ocean, see Geologic Notes 38, 42, 140, 142, 144, 146, 150, 151, 153, and 154.

chapter 4

Annealing the Parts

The foundation of the North Cascades is a complex mosaic of terranes whose early history includes deposition of sediments and volcanic rocks in oceans and volcanic arcs. Plate collisions brought the pieces together (accreted them), probably somewhere along the west side of North America by about 90 million years ago (Cretaceous). Only at this point can these assembled rocks properly be regarded as the block of crust today known as the North Cascades. After about 90 million years ago, the speed of movement lessened significantly, although even today pieces of the mosaic are still moving.

As demonstrated by the rocks of the Western Domain, the accretion process stacks up a considerable thickness of rock by thrusting one terrane over another. A consequence of such *overthrusting,* as it is called, is burial of rocks on the bottom of the thrust stack. If such rocks remain buried long enough, they heat up and become metamorphosed. The rocks of the Metamorphic Core Domain, the most rugged and mountainous part of the North Cascades, were created or modified during such a metamorphism at the bottom of a thrust stack, most of which began about 90 million years ago.

Much of the rock in the Metamorphic Core Domain began as magma intruded just before or during the metamorphism. These invading plutons supplied some of the heat for the recrystallization that occurred during the metamorphic episode. The remaining rock is schist of the Chelan Mountains and Nason terranes. Almost all these rocks exhibit metamorphic foliation and/or lineation (Chapter 2), and the alignment of minerals tells us that the rocks, both the original material of the terranes and the invading plutons, were being squeezed and stretched—deformed—as they were recrystallized.

For more information on foliation and lineation in metamorphic rocks, see Geologic Notes 7, 33, and especially the accompanying figures.

STITCHING PLUTONS

Because the plutons invaded most of the terranes that make up the North Cascades mosaic and extend from one to another like a staple, geologists call them *stitching plutons* (Plates 1, 5D). Most of the plutons do not really "stitch" the terranes together, but by invading rocks of many of the terranes all at about the same time, they show that the mosaic was by then assembled. (The terranes may not yet have been attached to the North American continent in northwest Washington—we discuss that part of the puzzle in Chapter 5.) Although stitching plutons are not found in the Western Domain at the present level of erosion (who knows what lies below?), some rocks there were also metamorphosed about the same time as those in the Metamorphic Core Domain, but to a much lesser degree.

Radiometric dating (Geologic Note 4) indicates that many stitching plutons are between about 65 and 90 million years old (Cretaceous). The origin of all this magma is not completely clear although many geologists think that it represents the roots of an old volcanic arc. Few of the ancient arc volcanoes or their immediate products have survived erosion, except perhaps in the Methow Domain, so we cannot be sure that all the plutons represent a single volcanic arc.

THE SKAGIT GNEISS COMPLEX

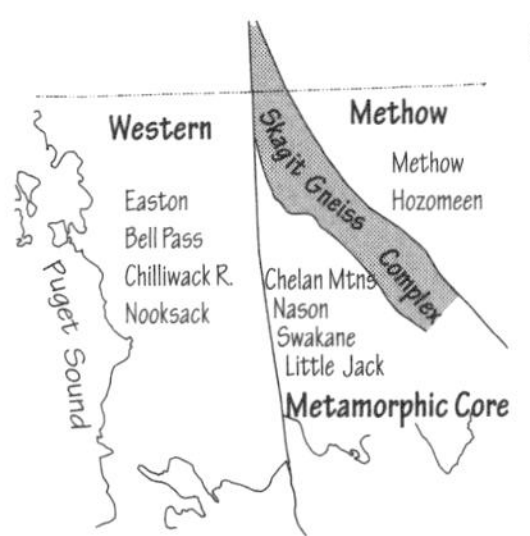

Domains and Terranes

The metamorphic Skagit Gneiss Complex is derived from the metamorphosed strata of the Chelan Mountains terrane by intrusion of many igneous sills and dikes, which were then metamorphosed. Metamorphism occurred about 90 million years ago and continued or renewed about 45 million years ago (Cretaceous and Eocene).

In the Metamorphic Core Domain, including much of the wilderness of North Cascades National Park, two kinds of gneiss, orthogneiss and banded gneiss, dominate the scene.

Large tracts of land began as igneous plutonic rock, now commonly with metamorphic foliation and/or lineation; such plutonic rocks are called *orthogneiss* (literally meaning "straight gneiss"), by which geologists mean gneiss derived from igneous plutonic rocks rather than from the thorough metamorphism of sedimentary or volcanic rocks. (Geologic Notes 4, 7, 28, 33, 111, 122, 124, 128).

Banded gneiss is made up of layers of granitic orthogneiss alternating with schist (Figure 23). The schist layers in the banded gneiss appear to be remnants of the Chelan Mountains terrane's Cascade River Schist and the Napeequa Schist (Chapter 3; Geologic Notes 19, 20, 22, 33, 65, 66, 71). The gneiss layers are mostly deformed igneous sills.

Geologists have lumped the banded gneisses and many of the more extensive orthogneiss bodies together under the name *Skagit Gneiss Complex* (see Plate 2). Most of the Skagit Gneiss Complex consists of what geologists call migmatite (see "Orthogneisses and Migmatites").

MORE METAMORPHISM

As if this invasion by stitching plutons and metamorphism were not confusing enough, beginning about

Orthogneisses and Migmatites

Orthogneiss of at least two different ages makes up much of the Skagit Gneiss Complex. Some magma intruded while rocks of the Metamorphic Core Domain were being squeezed and probably folded. While the magma crystallized or soon thereafter, the squeezing (or flattening) aligned the minerals and the rock became foliated orthogneiss. Twenty million years or so later, much of the squeezing had ceased. When new magma invaded, the newly crystallized granitic rocks developed much less foliation than their predecessors. But because they were subjected to an episode of stretching (Chapter 5), these younger granitic rocks exhibit strong lineation. Stretched minerals look like pencils or, on broken surfaces of the rock, scattered parallel dashes. (For more information on foliation and lineation, see Geologic Notes 7 and 33 and their accompanying figures.)

Most of the older orthogneiss has the composition of tonalite. Most of the younger orthogneiss has a composition closer to granite (see Note 95 and accompanying Figure 86). From isotopic dating of zircons, geologists know that the older magmas solidified about 65–90 million years ago (Cretaceous) and the younger ones about 45 million years ago (Eocene). The older orthogneiss bodies in the Skagit Gneiss Complex are stitching plutons. The younger ones embroider the already stitched-together terrane mosaic.

In many areas, the complex sequence of invading magma and deformation left a confusing mixture of rocks. Gneiss was cut by light-colored dikes or sills, which were then all squeezed and deformed. This deformation was followed by intrusion of still more dikes and further squeezing and stretching. Even more dikes may have intruded after that. Geologists call such rocks *migmatites* (from the Greek for mixed rocks), but they are hard pressed to describe them.

(To learn more about orthogneisses and migmatites, see Geologic Notes 7, 11, 20, 22, 116, 122, and 124.)

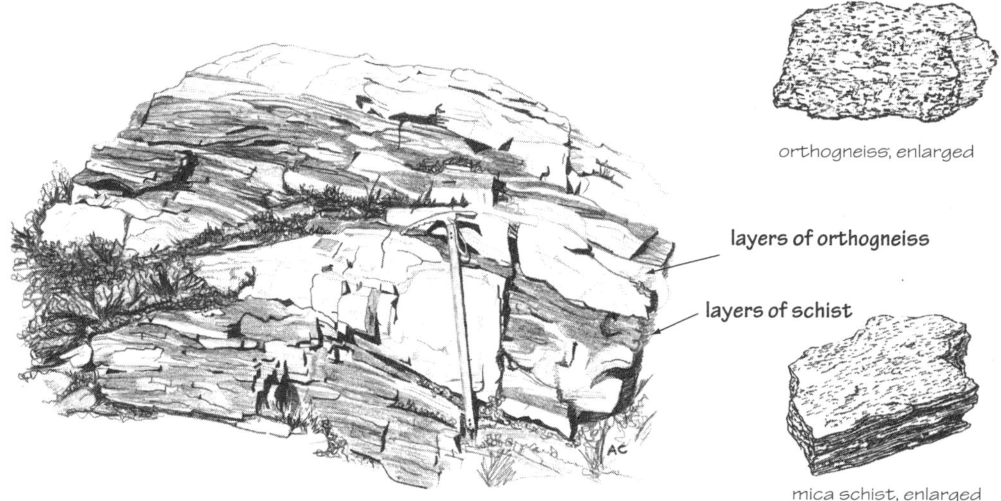

Figure 23. Banded gneiss.

10 million years later, when the Cascade Volcanic Arc awakened, arc-root plutons invaded the rocks of the North Cascades (Chapter 6). Near these hot magmas, intense heat caused the older schists and gneisses to recrystallize once again, but this time the rocks were not squeezed or stretched. The resulting *thermal*, or *contact, metamorphism* tended to obscure the foliation, lineation, and other effects of earlier metamorphic episodes (Plate 3A). The rocks baked in the zone around the plutons are harder and more resistant to erosion than their unbaked (or even half-baked) neighbors (Figure 24; Chapter 7).

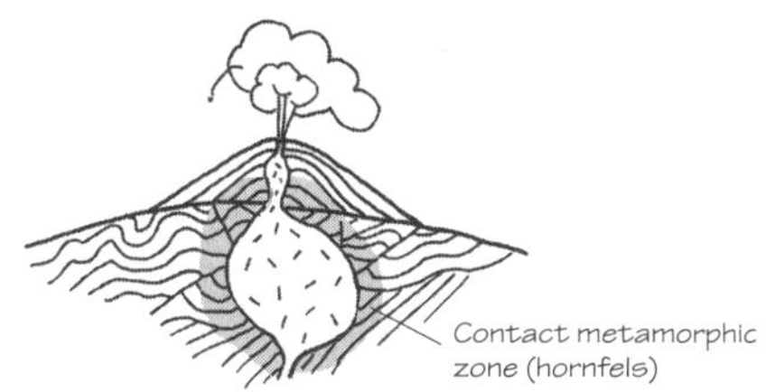

Figure 24. Cross section of small pluton beneath a volcano. Rocks around the pluton are baked.

chapter 5

Shifting the Pieces

About 90 million years ago, tectonic plate movements had brought together the terranes of the North Cascades and thickened the Earth's crust in the Pacific Northwest, profoundly metamorphosing its deeper part, as seen today in the Metamorphic Core Domain. Even after the terranes had arrived at roughly their current positions, the plates continued their slow rearrangement of crustal pieces. About 55 million years ago directions and rates of movement in the Pacific Northwest changed enough to cause local *extension* or stretching of the crustal rocks, and in places this extension continued until about 40 million years ago. Manifestations of the continued shifting of the pieces that now make up the North Cascade Range were northward drift of western crustal blocks relative to their eastern neighbors, sinking of some crustal blocks to form depressions where sediment accumulated, and—in conjunction with the extension that caused the sinking—intrusion of igneous plutons that caulked some of the cracks, that is, the faults created by the extension and northward drift.

NORTHWARD DRIFT

Three lines of evidence suggest that rocks in the North Cascades that are older than about 100 million years were transported northward from where they originally formed.

EVIDENCE FROM OLD FAULTS

The most direct evidence of the northward movement of rocks in the North Cascades can be seen in the distribution of the Nason terrane (see Plate 2; Figure 25). Two outcrop areas of the Nason terrane—one in Washington east of the Straight Creek Fault and the other in British Columbia, west of the Straight Creek Fault—are separated by at least 63 miles. Rocks of the two separate areas were born in similar places—if not the same place—and share similar metamorphic and deformational histories. Because of the similarities, geologists reason that rocks of the two areas once must have been adjacent. By making similar comparisons of rocks on opposite sides of many other faults in the North Cascades and Canada, geologists have gathered enough evidence to conclude that even if northward movement of crustal blocks on the west side of major faults has not been a hard rule over the last 100 million years or so, it has at least been a significant habit.

Such northward migration is still happening today all along the west coast of North America. It is most dramatic in central and southern California, where a large block of crust on the west side of the infamous San Andreas Fault is slowly moving northward, carrying Los Angeles towards San Francisco at a rate of about 1½ inches per year. In the North Cascades, movement along several faults—the Straight Creek Fault in particular—was at one time similar to that along the San Andreas Fault. Unlike along the San Andreas, movement has not occurred along the Straight Creek Fault—at least in Washington—for some 35 million years. At that time subduction began to generate the Cascade Volcanic

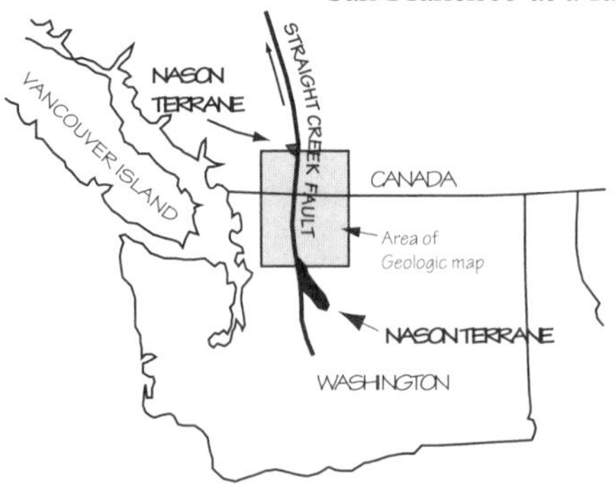

Figure 25. The Nason terrane has been offset by movement along the Straight Creek Fault.

Arc, described in Chapter 6. Collision between the oceanic plate and the North American plate had become more head-on, decreasing the tendency of parts of the continent to split off and move north. Geologists speculate that by this time, also, another fault farther west than the Straight Creek Fault, perhaps separating the Olympic Mountains from the North Cascades, had become the major fault accommodating northward drift of the west edge of the continent. Such northward drift continues today at a low rate and is a major cause of shallow earthquakes in the vicinity of Seattle.

EVIDENCE FROM TECTONIC PLATE MOTIONS

Additional evidence suggesting that the North Cascades originated far south of their current location comes from known patterns of plate movement. By studying the present position of all the Earth's tectonic plates, measuring their present rates of movement, and taking into account the geologic history of certain well-studied areas, geologists and geophysicists have been able to pretty well reconstruct tectonic plate movements over the last 150 million years. Most such reconstructions show a steady northward motion of plates in the Pacific Ocean relative to the North American plate over the last 80 million years. Even when and where the oceanic plate has subducted beneath the continental plate, rather than sliding past it, the constant northward push on the edge of North America tends to break off pieces of continent and move them northward. Probably part of the Nason terrane drifted north along the Straight Creek Fault into British Columbia during a later stage of this process.

Paleomagnetism: Finding a Rock's Place of Birth

Mountain travelers, especially those in a fog, become all too familiar with the needle of a compass as it swings back and forth until it lines up with the Earth's magnetic field, pointing towards the magnetic north pole. If the needle were positioned so that it could swing vertically, it would pivot until it reaches an angle that depends on the (magnetic) latitude; that is, on how far away the pole is. Such information might be useful to a hopelessly lost explorer. In the tropics the needle would remain nearly horizontal (Figure 26). At high northern latitudes, the compass needle would incline steeply towards the north. Sensitive compasses used by surveyors in the northern hemisphere are weighted on the south end of the needle to counteract this tendency. (Near the south pole, the north end of the needle tends to point steeply up, and southern-hemisphere surveyors move the weights to the north ends of their needles.)

Some iron-rich minerals are magnetic, and during their crystallization and cooling their magnetism aligns itself with the Earth's magnetic field. With suitable instruments, a paleomagnetist can measure this magnetism and read some rocks as if they were compasses that had recorded the direction (and inclination) of the Earth's field when the rocks formed.

Wholesale reversal of the Earth's magnetic field—swapping of north and south magnetic poles—causes alternating directions of magnetization in rocks of differing ages and gives rise to the magnetic stripes on the ocean floor (Chapter 2). But a closer reading of the magnetic alignment of certain rocks, described above, reveals smaller changes in the direction to the north magnetic pole. These changes are due to movement either of the pole or of the rocks themselves with respect to the pole. Both sorts of movement have happened.

In 1972 geologist and paleomagnetist Professor Myrl Beck and his students first reported evidence of large northward migration of stitching plutons. Since then, other paleomagnetists have measured other rocks about 100 million years old and found similar results. Measurements on arc-root plutons of the Cascade Volcanic Arc and on volcanic rocks associated with the arc indicate that these younger rocks formed more or less where they are today: terranes of the North Cascades were definitely in place at their present latitude by about 35 million years ago.

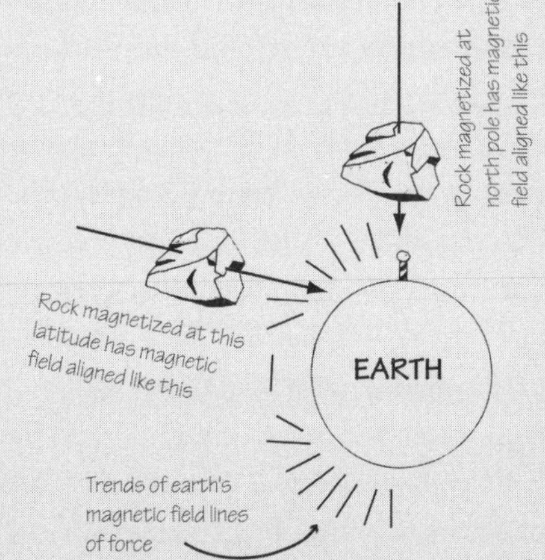

Figure 26. A rock's paleomagnetic alignment depends on the latitude where the rock formed.

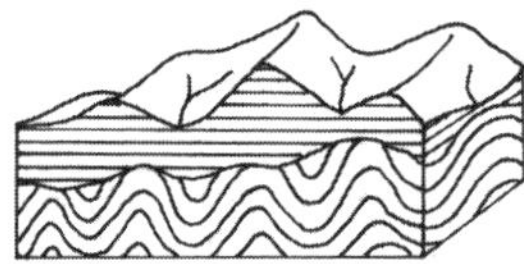

Figure 27. Unconformity between younger flat-lying beds deposited on top of older folded beds.

EVIDENCE FROM PALEOMAGNETISM

Finally, the most dramatic evidence of northward movement of the terranes that now make up the North Cascade mosaic is the magnetism recorded in some rocks. This paleomagnetism records, in a rough way, the distance from, and direction to, the Earth's magnetic pole when the rock formed (see "Paleomagnetism: Finding a Rock's Place of Birth"). If paleomagnetists—ordinary geologists call them paleomagicians—can find a rock that retains its original magnetism, and if they can determine how much it has tilted or twisted, then they can also determine at what latitude the rock was born. Several paleomagnetic studies have found that rocks now in the North Cascades and nearby in British Columbia moved north some 1,800 miles between 100 million years ago and about 50 million years ago. This conclusion is quite controversial, but each additional study—and paleomagnetists are very busy folk—supports the idea that North Cascade rocks formed far to the south of where they are now.

EXTENSION OF THE CRUST

On top of, and in strong contrast to, the strongly deformed and metamorphosed rocks found in the Western and Metamorphic Core Domains are younger, unmetamorphosed, sandstones and conglomerates. Most of these sedimentary rocks were deposited by streams or rivers flowing across the older rocks. Where young bedded rocks overlie eroded older rocks, geologists call the contact an *unconformity* (Figure 27). Such a break in the rocks generally indicates a corresponding time break in the geologic record. Simple uplift and erosion of sediments before deposition resumes may produce an unconformity—only a few million years of sedimentation missing—or the older beds may have been deeply buried, folded, metamorphosed, uplifted, eroded, and then submerged in the ocean for burial under younger sediments—a profound unconformity with maybe tens of millions of years of record missing.

Experts who study sedimentary rocks (sedimentologists) have concluded that the sandstone and conglomerate found unconformably on the older rocks of the North Cascades were deposited very rapidly, probably in down-dropping fault valleys or basins. Geologists believe that for the crust to create these depressions, it must have been stretching or extending (Figure 28). For this reason we call these sandstone and conglomerate accumulations *extensional deposits*. The extensional deposits erode rapidly and are only preserved in somewhat special circumstances (Chapter 7; Figure 35). In the area of this guide, extensional deposits of sandstone and conglomerate overlie older rocks southwest of the town of Glacier and in a few other places (see Plates 2, 8B; Geologic Note 60).

CAULKING THE CRACKS

Rocks in the deeper crust remained hot and plastic, both stretching and recrystallizing during extension and northward drift. At shallow levels the cool, brittle crust broke as it extended and the western pieces drifted northward (Figure 29). During the later phases of drifting and extension, about 45–50 million years ago (early Tertiary), plumes of granite magma rose into the deforming crust. One of these plumes of ascending magma became the Golden Horn batholith (see Plate 2, Geologic Notes 16, 138, 141), which caulks part of the Ross Lake Fault, one of the faults along which crustal blocks migrated north. During this magmatic phase the Skagit Gneiss Complex, still buried deeply enough to remain hot and pliable, was intruded by innumerable granite sills and dikes that were stretched into lineated orthogneiss (Chapter 4).

WHY EXTENSION?

The crust may extend because its strength decreases, and it collapses of its own weight (think of a softening block of butter), or because it is pulled apart. Or both. Geologists think that the 55- to 40-million-year-old (Eocene) extensional deposits of the North Cascades are mostly the result of the crust being pulled apart by the plate tectonic forces that for tens of millions of years previously had been moving the terranes northward. Softening may also have contributed to the extension because, as described in

Forming Depressions Along Faults

One way to get extension or stretching of the crust while blocks of crust are sliding past each other horizontally along large faults is to have irregularities in the fault planes that open up gaps in the crust, or to have the faults discontinuous but arranged in an *en echelon* pattern.

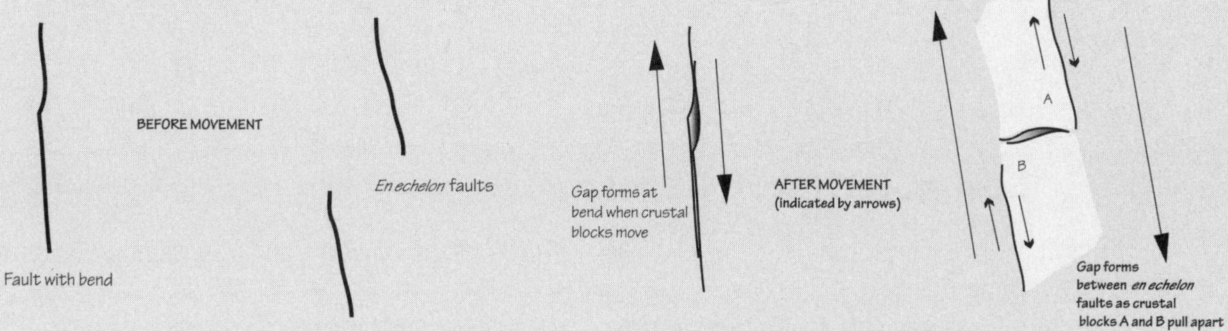

Figure 28. Diagrammatic maps showing how horizontal movement on faults can produce depressions that fill with sediment.

Chapter 4, a considerable mass of the accreted terranes was very hot, undergoing metamorphism, and lubricated by igneous plutons (which became the orthogneisses). A slight change in the pressures of the adjoining plates, and the whole structure would have collapsed.

The extensional event may have been initiated by rearrangement of plate motions in the adjacent Pacific Ocean. There is evidence that the rate of plate motion increased at this time, and perhaps a different plate came into contact with North America at the latitude of the North Cascades. The extensional event, most (but not all) of the northwards drift of western crustal blocks, and distinctive granitic magmatism all came to an end when plates in the Pacific Ocean again rearranged about 40 million years ago and the Cascade Volcanic Arc was born.

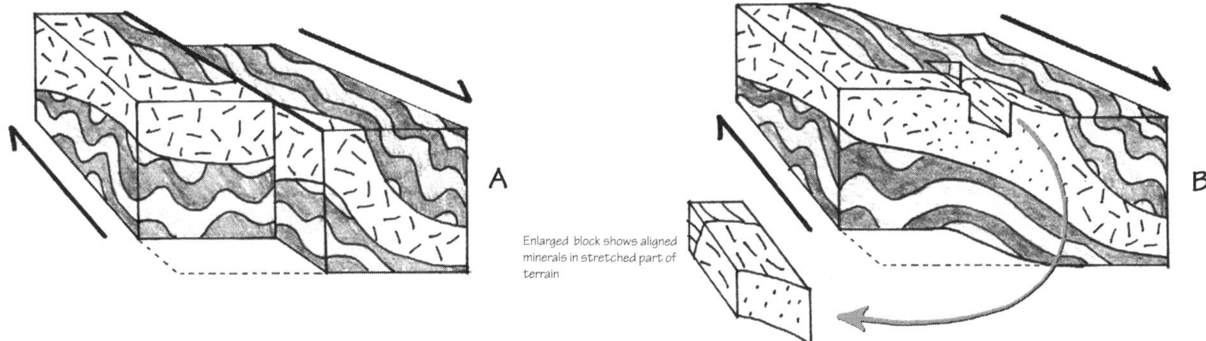

Figure 29. As the crust is deformed by northward drift, a granitic dike is (A) faulted near the surface (visible after some erosion) and (B) stretched like taffy at depth where the rocks are hot (visible only after much erosion).

chapter 6

From the Fiery Furnace

For people in the Pacific Northwest, the growth of the Cascade Volcanic Arc is the most dramatic and comprehensible act in the geologic drama presented in this book. The arc began erupting about 35 million years ago (Oligocene) and is still erupting. Lavas representing the earliest stage in the development of the Cascade Volcanic Arc mostly crop out south of the North Cascades proper, where uplift of the Cascade Range has been less, and a thicker blanket of Cascade Arc volcanic rocks has been preserved (Chapter 2; Figure 6). In the North Cascades, geologists have not yet identified with any certainty volcanic rocks as old as 35 million years, but remnants of the ancient arc's internal plumbing system persist in the form of plutons, which are the crystallized magma of chambers that once fed the early volcanoes.

ROOTS OF THE VOLCANOES

The greatest mass of exposed Cascade Arc plumbing is the Chilliwack batholith (Plate 5A), which makes up much of the northern part of North Cascades National Park and adjacent parts of British Columbia (Figure 30). The batholith is actually made up of numerous separate, mostly granitic, plutons, some large, some small, but all more or less contiguous. When the batholith was first named by Reginald Daly in the early twentieth century, the great range in ages of the individual arc-root plutons was not known. Individual plutons range in age from about 35 million years old to 2.5 million years old. Based on their ages, we have divided the plutons of the batholith, and in the North Cascades in general, into three families: the Cascade Pass family, the Snoqualmie family, and the Index family(see Plate 2). Each family includes a wide range of igneous rocks.

For more information on the families of the Chilliwack

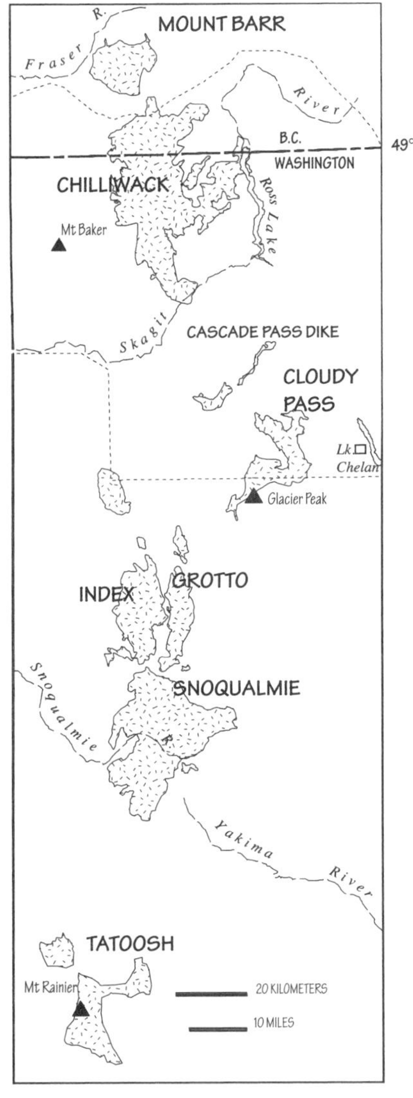

Figure 30. Cascade Arc batholiths and associated plutons in Washington and nearby British Columbia. Dashed lines show area covered by the geologic map (Plate 2).

Making Room for Magma

A backcountry hiker, standing in the upper Baker River Valley, looking up at the forbidding granitic cliffs of the Chilliwack batholith all around, and realizing that he or she stands in the middle of a once vast chamber of molten rock, might wonder (especially with prompting from this book) how all this granitic rock found room for itself in the rocks that were here before the magma arrived. No vast cavern to receive the rocks could have existed in the crust at the depths and pressures where this magma cooled. Geologists have proposed a number of scenarios, including having the invading magma crowd the invaded rocks aside, punch the roof rocks above the pluton up like a piston, or flow upward in the core of the pluton and down at the margins, carrying down and away blocks mined from the surrounding rocks. If the older rocks just melted to make the magma, not much new room would be necessary; but that does not seem to work. We know that, for the most part, not much of the surrounding country rock (at the level where we now see the batholith) was melted and incorporated into the plutons. The chemical compositions of the plutons do not reflect additions from the surrounding rocks. In the North Cascades, where the mechanics of intrusion have not been studied much, plutons probably made room for themselves mostly by the complex diapiric intrusion shown in Figure 31, but locally the other scenarios discussed may have played a part.

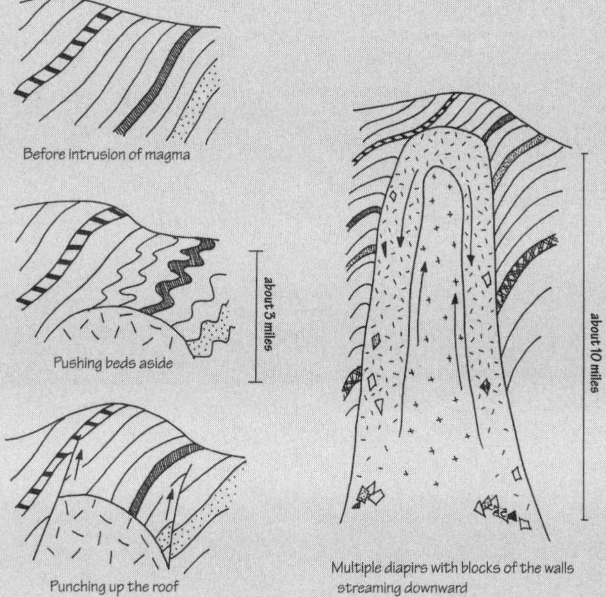

Figure 31. Cross-sectional views of various scenarios proposed for how intruding magma was accommodated within existing rock (sketch of multiple diapirs after Paterson, Fowler, and Miller, 1996).

batholith, see "Families of the Arc-Root Plutons."

The older rocks invaded by all this magma were affected by the heat. Around the plutons of the batholith, the older rocks recrystallized (Figure 24). This contact metamorphism (Chapter 4) produced a fine mesh of interlocking crystals in the old rocks, generally strengthening them and making them more resistant to erosion. Where the recrystallization was intense, the rocks took on a new appearance—dark, dense, and hard. Geologists call such rocks *hornfels,* that is, horn rock. Many rugged peaks in the North Cascades owe their prominence to this baking. The rocks holding up many such North Cascade giants as Mount Shuksan, Mount Redoubt, Mount Challenger, and Hozomeen Mountain are all partly recrystallized by plutons of the nearby and underlying Chilliwack batholith.

TRACES OF THE LAST ARC

Very little is left of the old Cascade Arc volcanoes fueled by the deep magma chambers now represented by

Making Granite and Its Relatives

The different kinds of plutons—granite, granodiorite, tonalite, gabbro—encountered in the North Cascades, and mentioned throughout this book, acquired their different compositions mostly through two very different but related processes: *magmatic differentiation* and *differential melting*.

Magmatic Differentiation

When magma begins to crystallize, the first crystals to form are commonly those of minerals richer in iron, magnesium, and calcium. These are dark minerals, such as hornblende and pyroxene, and calcium-rich plagioclase feldspar. As a result, the remaining melt has proportionately less iron, magnesium, and calcium, and relatively more elements such as sodium, aluminum, and silicon. If this melt for some reason drains away from the early-formed crystals, it will, when it crystallizes, produce a lighter-colored rock that contains more quartz, potassium- and sodium-rich feldspar, and biotite (Figure 32).

Magmatic differentiation has been demonstrated to work in the laboratory and careful chemical studies of granitic rocks substantiate it as well. Geologists believe that this process can produce all the different kinds of plutons of similar age occurring in a local area. In some plutons, a gradation from dark rocks rich in hornblende and pyroxene to a core of light-colored rock indicates that the liquid remaining after the early crystals formed became concentrated in the center of the mass. By looking at the chemical changes in suites of granitic rocks and volcanic rocks, geologists surmise that the usual beginning melt has the composition of basalt not unlike that of the ocean floor. Magma may leak to the surface at any stage in the differentiation process, giving rise to a corresponding variety of volcanic rocks.

If basaltic magma crystallizes slowly but without differentiation, the resulting rock will be a *gabbro* (the plutonic equivalent of basalt; see Glossary and Figure 112). We find such gabbros in several places in the Chilliwack batholith. One is on Mount Sefrit, west of Ruth Creek, and another is near Copper Mountain above the Chilliwack River. The basaltic magma that produced these dark rocks came up from great depths without pausing to differentiate. However, most of the root plutons of the Cascade Volcanic Arc and the volcanic remnants are differentiated somewhat, to produce tonalite and granodiorite and their volcanic equivalent, dacite. Only a few

the Chilliwack batholith. Only where these older volcanic deposits have been down-dropped by faulting have they been preserved from total erosion. Remnants of these deposits occur at Silver Creek (west of Ross Lake), Big Bosom Buttes Caldera, Pioneer Ridge, Hannegan Pass Caldera, and Kulshan Caldera.

For more information on the ancient remnants, see "Old Volcanoes."

REMNANTS OF YOUNGER VOLCANOES

After Kulshan Caldera formed more than a million years ago, considerable amounts of lava erupted from various vents around what is today Mount Baker, which did not yet exist. The Black Buttes, on the west side of Mount Baker, are the remnants of an older cone that grew about 500,000 years ago. Its lavas cap nearby Heliotrope and Marmot Ridges. Flows from Black Butte volcano also make up Forest Divide on the east side of Mount Baker. Volcanologist Wes Hildreth estimates that the Black Butte volcano was roughly twice as big as the present-day Mount Baker volcano. Most of it was eroded away before Mount Baker came to life.

Another prominent pile of lavas makes up Table Mountain, underlies parts of the Mount Baker Ski Area, and forms much of Ptarmigan Ridge. These lavas are about 310,000 years old, and their thickness suggests that they flowed down ancient valleys.

THE MOUNT BAKER VOLCANO

Mount Baker is one of the youngest members of the Cascade Volcanic Arc. It is probably less than 30,000 years old and has not been eroded enough to expose its granitic roots. The most conspicuous young lavas associated with the volcano flowed down Sulphur Creek less than 10,000 years ago, erupting from a small cinder cone near Schriebers Meadow. An explosive eruption

Cascade arc plutons differentiated to granite and its volcanic equivalent, rhyolite. Volcanic andesite, the result of only slight differentiation, is present in the Cascade Arc, but its plutonic equivalent, diorite, is rare, a discrepancy possibly due to lack of data.

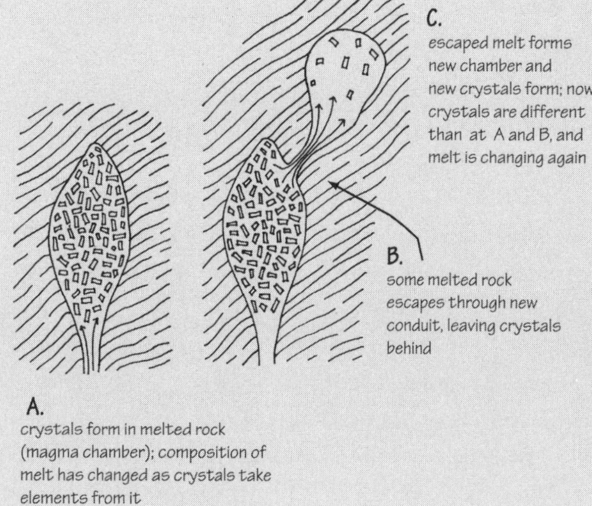
C.
escaped melt forms
new chamber and
new crystals form; now
crystals are different
than at A and B, and
melt is changing again
B.
some melted rock
escapes through new
conduit, leaving crystals
behind
A.
crystals form in melted rock
(magma chamber); composition of
melt has changed as crystals take
elements from it

Figure 32. Simplified sketch of magmatic differentiation as seen in cross section of magma chamber.

Differential Melting

Differential melting is the reverse of magmatic differentiation. In a subduction zone, the amount of water rising from a descending oceanic plate into the overlying mantle and lower crust varies a great deal depending on how much is in the subducted plate in the first place. Larger amounts of water will cause hot rocks to melt sooner and at a lower temperature than they would if permeated with lesser amounts of water. The resulting melted material will therefore have more silica—the reverse of the magmatic differentiation process described above—because the magma is not hot enough to melt all the constituents of the iron-, magnesium-, and calcium-rich minerals. If this early-formed melt is separated from the remaining unmelted rock, a magma with composition different from the unmelted material will result. Professor Jeffrey Tepper, who, along with his colleagues, has studied the Chilliwack batholith and other North Cascade arc-root plutons more than most, believes that the amount of water released from the subducting plate has greatly influenced the composition of the plutons in the batholith.

(For more information on the composition of igneous plutons, see Geologic Notes 92, 93, 95, and 138.)

Families of the Arc-Root Plutons

Family	Age
Cascade Pass family	Less than 20 million years (m.y.) old

Comment

Named for the big dike at Cascade Pass (see Geologic Notes 64 and 69; Plate 2). The youngest member of the family so far discovered is the Lake Ann stock (see Geologic Notes 80 and 92), about 2.5 million years old.

Family	Age
Snoqualmie family	22 to 28 m.y. old

Comment

Named for the Snoqualmie batholith, exposed along Interstate 90 at Snoqualmie Pass, east of Seattle.

Family	Age
Index family	29 to 35 m.y. old

Comment

Named for a pluton making up part of Mount Index, east of Seattle on U.S. Highway 2.

Old Volcanoes

Volcano	Age
Kulshan Caldera	1.1 million years (m.y.) old

Comment

The Kulshan Caldera is an elliptical depression, more than 3,000 feet deep, filled with violently erupted volcanic ash. According to volcanologist Wes Hildreth, the hole making the caldera formed as magma erupted, and overlying rocks subsided into the magma chamber. Much of the erupted material fell right back into the hole.

Volcano	Age
Hannegan Caldera	About 4.4 m.y. old

Comment

Best viewed by hiking to Hannegan Peak, north of Hannegan Pass. Like Kulshan Caldera, formed in a violent eruption.

Volcano	Age
Big Bosom Buttes	About 30 m.y. old

Comment

Can be viewed from Twin Lakes and vicinity. The preserved volcanic breccias and tuffs erupted onto eroded hills carved from granitic rocks of the Chilliwack batholith. The bottom layers of breccia are rich in fragments of the older granitic rocks. The volcanic rocks were later intruded by younger granitic magma of the batholith.

Volcano	Age
Volcanic rocks of Pioneer Ridge	25 to 32 m.y. old

Comment

On Pioneer Ridge, south of the Picket Range. Also erupted onto eroded Chilliwack batholith.

Volcano	Age
Volcanic rocks of Mount Rahm	25 to 45 m.y. old

Comment

On Silver Creek, west of Ross Lake, and in Canada. Might represent some of the earliest volcanoes of the Cascade Volcanic Arc.

(For more information on the old Cascade Arc volcanoes, see Geologic Notes 52, 77, 78, 79, 88, and 94.)

of steam blew out of Sherman crater, near the summit of Mount Baker, in 1843, and steam still hisses into the sky from vents around the rim. We have every reason to expect future eruptions from Mount Baker.

Of more than philosophical interest is the question of whether today is a time of arc quiescence. Are the growth and violent activity of the Cascade volcanoes more subdued now than they have been in much of the recent geologic past? In describing the rocks generated by the arc, we tend to be overawed by the amounts of volcanic material making up the mountainsides. Keeping in mind that all the arc rocks mentioned, rocks of both the volcanoes and their plutonic roots, were generated over a time span of 35 million years, the amount produced in any hundred-year period or even thousand-year period is not so much. The scene we see today, with only a few high Cascade volcanoes, such as Mount St. Helens or Mount Baker, erupting every few hundred years or so, may not be far from the scene along the Cascade Volcanic Arc at any given time in the recent geologic past. Man's span of attention almost fits between the catastrophic but spaced events that build a volcanic arc.

For more information on Mount Baker, see Geologic Notes 53, 55, 56, and 59.

chapter 7

The Constant Levelers

In spite of the romantic notion that mountains are thrust up into the sky, or even geologists' chatter about mountain building, most mountain scenery is the product of erosion gradually reducing the mountains back to sea level. To produce mountains, a sizable bit of the Earth's crust must be elevated significantly above sea level. As this happens, the main agents of erosion—gravity, running water, and moving ice—get to work, carving the upraised crust into interesting topography. The overall driving force is gravity. In the long run, gravity constantly reduces elevated parts of the Earth's crust, returning the planet to a perfect spheroid as it was at birth.

When looking back in geologic time at the erosional history of the North Cascades, geologists have to decide at what point in the geologic past they want to call this collection of rocks the North Cascades. Much of the range is made up of exotic terranes that probably did not evolve on the same spot on the Earth as the present North Cascades. Many of the pieces had their own erosional histories before they became part of the North Cascade mosaic, and that early history is obscure. But if geologists confine their view to some time since the earliest Tertiary, that is, about 65 million years ago, they can speculatively recreate the North Cascade scene and ponder its erosional history.

By earliest Tertiary time, most of the terranes of the North Cascades were in place, and stacking by overthrusting and general squeezing of the terranes had thickened the crust significantly. A thick crust tends to float high on the denser mantle, like an iceberg in the sea, so it is safe to assume that by this time the mountains were elevated and erosion had begun its work. Even today, as erosion removes the top of the stack, the mountain root continues to rise.

THE WORK OF RUNNING WATER

To understand the role of stream and river erosion in shaping the North Cascades, geologists look to the events creating the major drainages of the region, such as the Columbia River, which cuts across the present-day Cascade Range. The river's immediate ancestor probably developed from streams running off an elevated block of crust in Washington and British Columbia dating from the uplift of land underlain by the thrust-thickened crust, probably sometime between 90 and 50 million years ago (middle Cretaceous to early Tertiary). Tributaries to the Columbia River and other lesser rivers to the north began etching out the ancestral North Cascade mountains, probably in a linear trellis pattern reflecting the major northwest-southeast alignment of the major rock units (see "How the Rivers Work").

Not long after (geologically speaking, of course) the ancestral Columbia River was established, probably about 60 million years ago (in the earliest Tertiary), tensional forces in the crust broke the region into fault blocks. Some blocks rose high enough to become mountains, some sank low to become basins. Streams began to fill the basins with sediments mostly eroded from nearby highlands. Professor J. Hoover Mackin, an ardent fan of Pacific Northwest geology, imagined a scene not unlike the Basin and Range region of Nevada and Utah today, with upraised fault-block mountains, exposing old rocks, surrounded by alluvial basins, slowly filling with sand and mud (Figure 35). Mackin called this collection of sunken basins and uplifted blocks the Calkins Range,

How the Rivers Work

When a piece of the Earth's crust first emerges from the sea and rain begins to fall on it, the water running off begins to erode it. As the first gullies develop into channels, the drainage of this new land gets organized by an interplay of three principal factors: differential erosion, stream capture, and base level.

Differential Erosion

As water runs downhill, it erodes channels that become organized into a drainage pattern. As running water deepens its channel, it encounters rocks of different hardness and adjusts its channel in response. Soft layers erode faster than hard layers. This simple statement is an important principle of erosion that applies to almost every natural scene. Eventually, hard layers or other hard parts of rocks stand out in bold relief, and softer layers or parts retreat into swales, gullies, and valleys. Geologists call the process *differential erosion.*

For much of their history, major rivers may be able to maintain their courses across rocks of different hardness. Differential erosion will produce rapids or falls where the rivers cross harder rocks, but for the most part the rivers will maintain their general course. Small side streams, on the other hand, will be more quickly influenced and their courses will adjust to follow weaker layers of rock (Figure 33). A look at the geologic map (Plate 2) will show that many rock bodies in the Metamorphic Core and Methow Domains in particular are aligned northwest-southeast. Structures such as bedding and metamorphic foliation are aligned in the same direction. This structural trend has influenced most major streams and rivers tributary to the Columbia River in the North Cascade region.

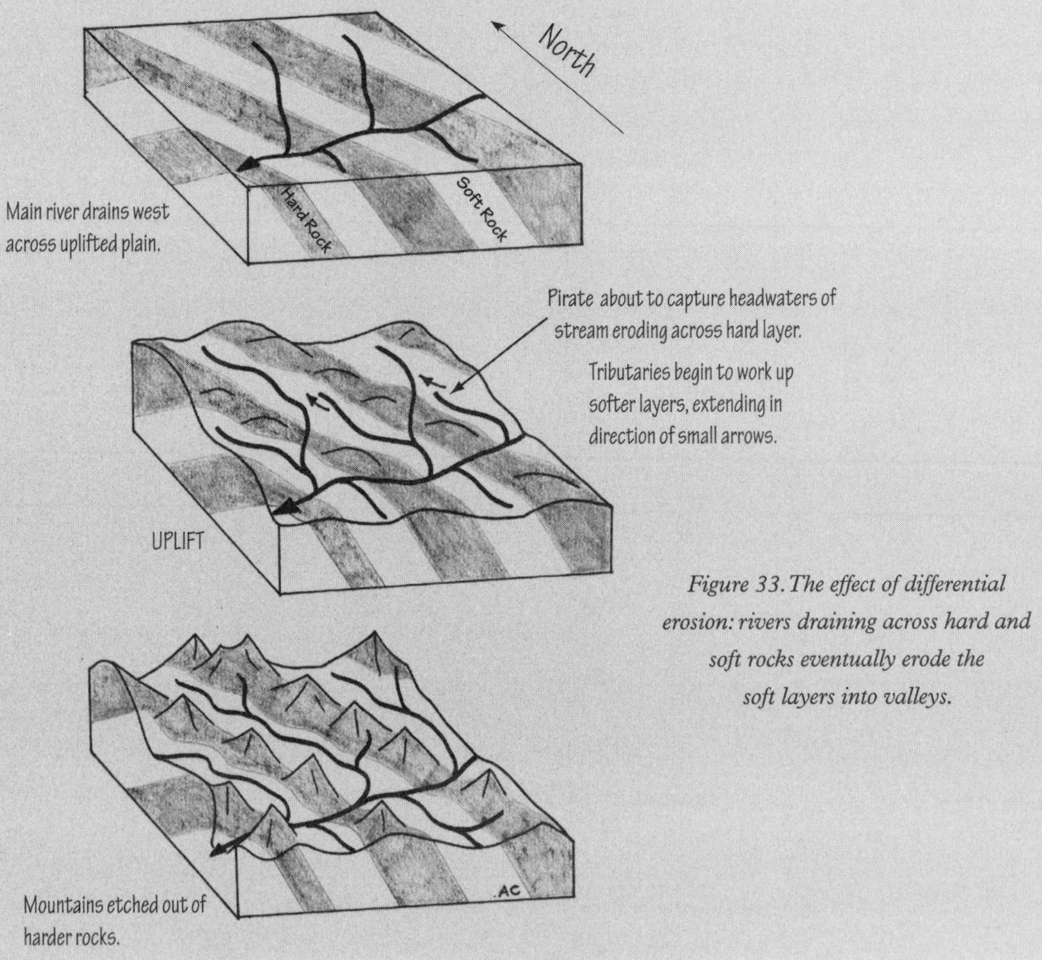

Figure 33. The effect of differential erosion: rivers draining across hard and soft rocks eventually erode the soft layers into valleys.

Rock hardness is an important factor in erosion, but a stream's ability to cut depends also on how much debris it carries—that is, the pebbles, sand, and silt which are its cutting tools—and its gradient, which controls its velocity.

Stream Capture

Stream capture, or piracy, is the process whereby a stream that is easily deepening its valley in soft rock can cut headward across a drainage divide to capture a neighboring stream that is working away slowly in harder rock. Examples of pure stream capture are hard to find in the North Cascades because of the overriding disruption of drainages by the growth and retreat of the Cordilleran Ice Sheet, but the overall parallel conformity of most North Cascade streams and rivers to the northwest-southeast-trending structural grain indicates stream piracy has ruled for millions of years. Figure 33, representing an idealized situation, illustrates the process.

Base Level

A fast-flowing stream with many rapids obviously has more energy for erosion than a slow, placid, meandering stream. Therefore, a young stream on a newly uplifted mountain block will erode rapidly, cutting a deep valley, and the lower parts of the stream or river, having more water and plenty of cutting tools, will cut deeper faster than the headwaters. Soon, the lower reaches of the stream will have a low gradient that approaches the flatness of the sea or lake into which it flows. This lower end point is called the stream's *base level.*

The profile of the stream moves towards a concave curve, flat at the lower end, curling up more steeply at the upper end (Figure 34). Eventually, if no geologic events interrupt the downward cutting of the stream, it reaches a state of pseudo-equilibrium, where the upper reaches are cutting very, very slowly because the rock resistance almost matches the energy of the falling water. At the lower end, the stream cannot cut deeper than its base level. Therefore, its energy goes into cutting its banks sideways. As a result, the stream meanders, gradually widening its valley.

Presumably, if the process continued long enough, the land would be reduced to sea level. Usually, however, the cycle is interrupted many times by geologic events, especially in mountains like the North Cascades. Rockslides may dam streams; uplift of the land gives streams new energy; glaciers grow and carve valleys into different profiles, or divert streams to a new locales; and volcanoes grow and change the landscape altogether.

Any change in the base level will propagate up the stream in some manner to affect its flow. A rockslide damming a river makes a new, temporary base level for the upper part of the stream. Naturally, the stream deposits all the debris it carries in the lake that forms. Eventually the stream spills over the top of the dam and, falling rapidly down the face, begins to erode it away. All lakes are temporary interruptions in a stream's course. We must enjoy them while we can.

For more information on how rivers work, see Geologic Notes 6, 24, 36, 37, 50, 74, 120, and 139.

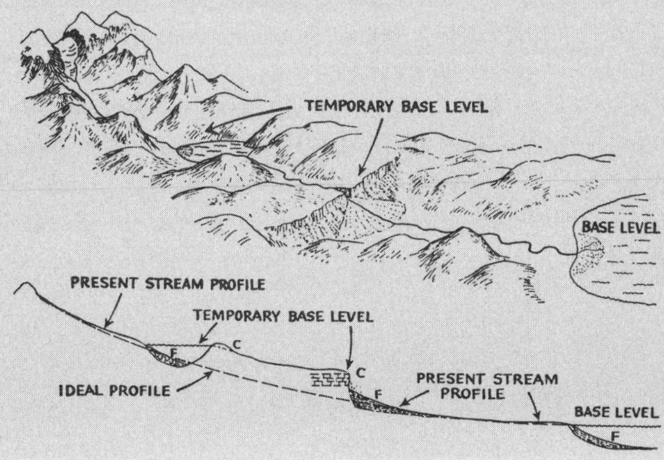

Figure 34. Graded stream profile and base levels. Stream is cutting at C, depositing sediment at F. Dashed line is ideal profile (from Manning, 1967).

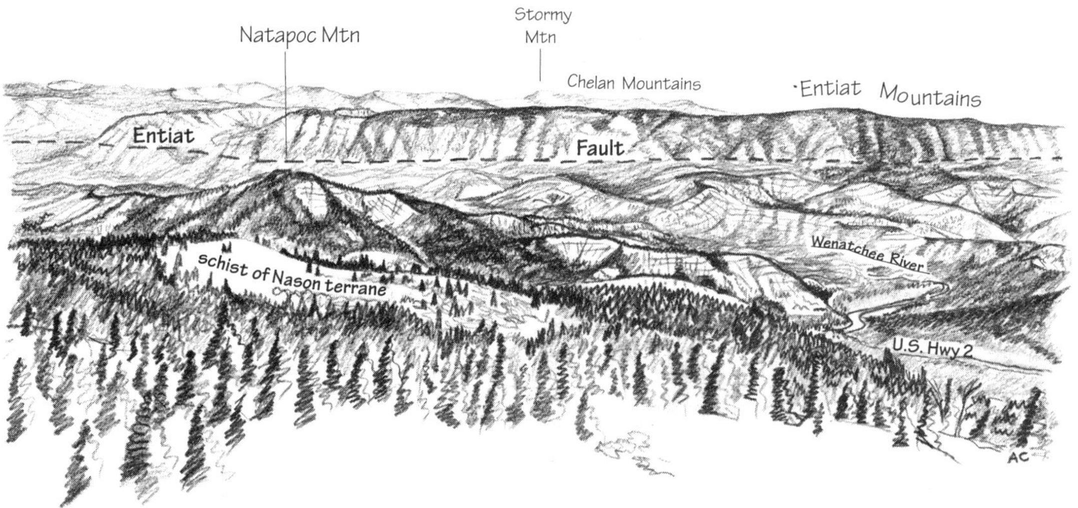

Figure 35. View to the east across the Chiwaukum Graben, a down-dropped block of river-deposited sandstone and conglomerate, which first formed as a structure in the Calkins Range. Visible strata in the graben show the folded beds of sandstone. The more easily eroded sedimentary rocks form the low hills between higher ridges of resistant gneiss and schist such as are seen here in the higher Entiat Mountains. The west-bounding fault is hidden behind the foreground ridge of schist.

after Frank Calkins, an early North Cascades geologist.

Although much of the material deposited in the basins came from nearby highlands, some of the sand grains in the sediment came from farther east, supporting the idea that rivers, like the Columbia, had already established their courses through the uplifted terrain. Not only were basins forming as the crust was stretching, but blocks of crust to the west were generally moving north, relative to blocks to the east, along major faults such as the Straight Creek Fault. Volcanoes must have erupted above some of the plutons associated with the Eocene extensional event, and some of the stream patterns of the North Cascades may well be relicts from rivers draining radially off these volcanoes.

Whatever drainage pattern was established, it was profoundly altered about 35 million years ago (Oligocene) by the renewed volcanic activity of the Cascade Volcanic Arc. Although only remnants of these ancient volcanoes exist here and there in the North Cascades (Chapter 6), the whole range during this period was probably blanketed by lavas and breccias, just as it still is today in southern Washington (Figure 6).

The growth of these volcanoes must have diverted numerous streams and established a mostly new drainage system, one reflected in the position of the major drainage divide today. This divide extends roughly north-south through North Cascades National Park, from Mount Redoubt on the north to Damnation Peak on the south, and separates streams that flow west into the Chilliwack, Baker, and lower Skagit Rivers from east-flowing streams that join the upper Skagit. This pattern may have been established as streams flowed down the slopes of the arc volcanoes that once erupted above the Chilliwack batholith. However, as this volcanic cover was eroded off, differential erosion worked its magic in readjusting the streams and rivers to the hardness and softness of the rocks beneath the volcanic cover (see "How the Rivers Work"). Much of the drainage may have found itself back in the old patterns it had before the volcanic interruption. Creeks once again widened fault valleys, and hard metamorphic and granitic rocks were left as lofty crags.

As the range continued to rise, the volcanic rocks were stripped away, except in places where they had been faulted down into the older rocks and are still preserved. Some are preserved because they were metamorphosed to hard rock by nearby intrusion of arc-root plutons. Mount Spickard and Ruth Mountain are examples of rugged peaks underlain by recrystallized (hornfelsic) volcanic rocks.

The overall drainage pattern we see today, with a few spectacular exceptions, was well established by the time extensive glaciers began to form about 1.6 million years ago (Pleistocene). Pleistocene glaciers, however, put the final touches on almost every scene. The evidence of glacier erosion or deposition is everywhere.

THE WORK OF MOVING ICE

Visitors to the North Cascades can easily perceive the role of glaciers in creating the mountain scene. Glaciers have left their marks everywhere, either in erosional landforms or in glacial deposits. Views of glaciers and their deposits abound. Backcountry hikers and climbers learn the joys and hazards of traveling on glacier ice. In general, to see the larger glaciers, the visitor must take to the trails and off-trail routes, but a few good distant views can be found with minimal effort. The Diablo Lake Overlook on Highway 20 offers views of the glaciers in the Colonial Peak group to the south. Mount Shuksan glaciers are on grand display from the road to the Mount Baker Ski Area (State Route 542) and from Artist Point. Hikes to Mount Baker from Schriebers Meadow or Glacier Creek afford even closer views of active glaciers.

FORMATION OF GLACIERS

Glaciers form wherever winter snowfall exceeds summer melt for enough years to accumulate a thick mass of snow. As the snow deepens, the bottom layers metamorphose into dense ice. The weight of overlying ice and snow causes the lowermost ice to flow, and a glacier is born. As the glacier grows, it extends farther and farther downhill. In the summertime, all the snow may melt off the lower end of the glacier, leaving bare metamorphic ice, some of which may also melt. At higher elevations, snow accumulates, transforms to ice, and creeps slowly down to the zone of melting. The boundary between the upper part of the glacier, where snow accumulates from year to year, and the lower part, where melting exceeds accumulation, is called the *firn line*. If the overall rate of melting exceeds the rate of accumulation, the lower end of the glacier retreats; if accumulation dominates, the lower end of the glacier advances down the valley. But the ice and snow within the glacier always are moving downslope.

HOW A GLACIER SCULPTS AND WHAT IT LEAVES BEHIND

A glacier sculpts the land in ways different from running water. A river occupies a small U-shaped channel in the bottom of a V-shaped valley. A glacier carves the whole valley into a U-shape (Figure 36). The upper end of a river is a small stream in a small swale that merges with the drainage divide. The head of a glacier-carved valley is a steep-walled bowl, or *cirque*. Glaciers are not as easily influenced by rock hardness as are rivers. They tend to bulldoze ahead, straightening out twisty river valleys.

Glaciers move very slowly, but can carry large amounts of rock debris. Individual blocks can be huge compared to the largest rocks moved by running water. When glaciers melt, they drop their loads as masses of unsorted debris, called *till*, in piles of various shapes, called *moraines*. Commonly, at the terminus, or snout, of a glacier, till forms a horseshoe-shaped moraine, with the open side facing upvalley (Figure 37). This *end moraine* forms as the glacier brings debris downvalley to the melting snout. The debris piles up as if delivered from a conveyor belt, albeit a slow one. Similarly, as a glacier moves downvalley, it also deposits till along its margins, forming sinuous ridges that parallel the glacier's flow. These ridges are called *lateral moraines*. Running water from the melting glacier can sort the unsorted till sediment into more uniform layers or beds of sand, mud, and/or gravel. Till is commonly mixed with this sorted material from the melt streams.

For more information on glacial features, see Geologic Notes 18, 21, 31, 47, 55, 62, 82, 83, 117, 118, 121, 131, and 134.

SCENERY BORN OF ICE

North Cascade valleys are U-shaped, with steep, cliffy walls and broad flat floors. The valley shape reflects erosion by long tongues of ice descending from peaks at the valley heads. Steep steps commonly occur at the junction of two or more major valleys where the combined action of the merged ice streams eroded faster and deeper.

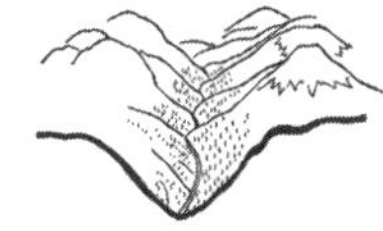

Figure 36. Profiles of V- and U-shaped valleys

Hikers nearing the heads of North Cascade valleys

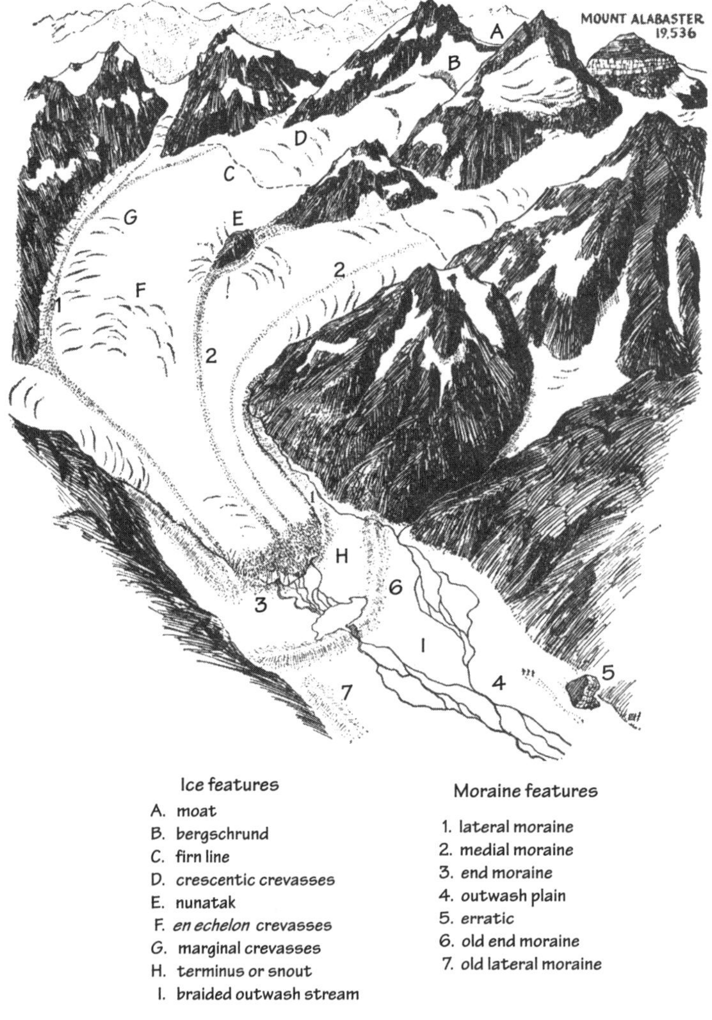

Figure 37. Glacier and moraine nomenclature (modified from Manning, 1967).

commonly encounter double cirques, making for two stiff climbs before reaching the high ridge tops. The lower cirque wall, a steep step in the valley, represents the head of an earlier great glacier, one that filled the valley during the Pleistocene. The upper cirque step, commonly still bearing a glacier, forms the head of the valley, and represents a new, higher, and smaller glacial bite into the ridge. Such multiple cirque steps are encountered on the east side of Cascade Pass, where the trail climbs from the lower valley to Pelton Basin (one) and then to the pass itself (two), and also on Bacon Creek below Berdeen Lake and on Railroad Creek at Crown Point Falls.

During the latest Ice Age, about 25,000 to 13,000 years ago (near the end of the Pleistocene), and probably during previous ice ages, much of the North Cascades was covered by glacial ice. During that time of colder climate, the local peaks grew their own glaciers, but in the higher mountains of British Columbia a vast glacier, the Cordilleran Ice Sheet, was growing and advancing south into Washington State. Eventually, Cordilleran ice filled the Puget Lowland, and its surface rose to merge with ice originating in the North Cascades. Only the highest peaks, generally those above 6,000 to 7,000 feet, stuck up above the ice surface. This mass of slowly creeping ice smoothed and rounded off the lower peaks, but much of today's scene was established by the melting of the ice sheet, not by its scraping. Huge amounts of water from the melting ice poured through North Cascade valleys. Lakes formed, deposits grew; lakes drained, and the same deposits were partially washed away. Some rivers that originally drained north to the Fraser River in British Columbia could no longer do so while the ice sheet remained. As a result, the water from the melting ice spilled south, cutting new and deeper canyons that held the rivers' courses even after all the ice was gone. Drainage reversal was widespread in the North Cascades.

The Skagit River is the most prominent permanent reversal. Elsewhere, meltwater carved deep notches, such as Chilliwack Pass at the head of the Chilliwack River, but failed to cut deeply enough to reverse much of the drainage once the ice had melted away. Actually, this story is more complex, for the drainage reversals could have developed in any of several earlier continental glaciations. Geologists find evidence of at least six major advances of ice into the Puget Lowland over the past 2 million years. Evidence for these earlier glaciations and possible drainage reversals is sparse in the mountains, where it has been obliterated by subsequent glaciations.

For more information on drainage reversal, see Geologic Notes 5, 6, 23, 35, 97, 139, and 149.

After the late Pleistocene retreat of the glaciers, the climate warmed to such an extent that most if not all the ice in the mountains disappeared. Then about 5,000 years ago, glaciers grew again. The glaciers we see today are remnants of this latest episode. Sometime between the thirteenth and nineteenth centuries, during the so-called Little Ice Age, small glaciers high in the Alps of Europe, the Himalayas, and many other mountain ranges including the North Cascades, advanced again downvalley one or more times and left conspicuous terminal moraines. If there were other glacier advances in the last 13,000 years, their tracks have been covered by these latest advances.

THE WORK OF GRAVITY

Erosion is ultimately an effect of gravity, that force which tends to keep the Earth a spheroid in spite of internal upheavals. Running water, ice, and hikers cutting switchbacks are merely intermediaries, helping move material downhill. In the mountains, gravity also works directly. Hillsides are moving slowly downhill most of the time in a process called *creep*. If a mass of material breaks away to fall downhill more rapidly it is a landslide or rockfall. Creep and landslides eventually turn the steep-walled gorges sawed by rivers into V-shaped valleys and hide the U-shape of glacial channels.

In steep-walled, glacier-carved valleys, landslides can be major actors in the erosional scene. Some move slowly, creeping downhill at a few inches or yards per year. Others are catastrophic. An especially large catastrophic landslide filled the valley bottom near Glacier on the North Fork of the Nooksack River about 2,400 years ago. Another large catastrophic landslide on the West Fork of the Pasayten River has not been dated directly but may have come down onto the last glacier to occupy the valley about 14,000 years ago. A landslide dammed the Skagit River north of Marblemount for about 1,500 years, long enough for a considerable amount of sediment to accumulate in the lake that formed. Radiocarbon ages of drowned trees in the lake indicate the mountainside collapsed about 8,400 years ago. Geologists do not know for sure what triggered these particularly large slides, but they suspect that earthquakes may have been the cause (Geologic Notes 3, 25, 73, 81, 152).

The smoother and lower the hills get, the slower the descent of rock and debris to the rivers and the slower the removal of the material to the sea. But gravity never tires and, given enough time, it will smooth even the boldest peaks of the North Cascades into low, rolling hills.

FEELING THE MOUNTAINS' PULSE

A traveler in the North Cascades, on the highway or along the trails, will soon get a feeling for nature's changing and evolving ways. Roaring cascades, fresh rockfalls, uprooted 500-year-old trees, all indicate the slow-motion change of the mountain scene. Many of the significant changes in the scenery come in bursts during great storms, or earthquakes, or volcanic eruptions. We brief visitors can see the evidence of these more recent changes all around us in the mountains, and if we look carefully at the rocks, we can also see the evidence of similar changes in the geologic past. A thoughtful visit to the North Cascades may let us all feel the slow but ever present pulse of the mountains and in them the pulse of the Earth itself, the planet we call home.

PART II

GEOLOGIC NOTES FOR POINTS OF INTEREST

The Geologic Notes of Part II are arranged by major rivers, beginning with the Skagit River and ending with the Pasayten River. Points of geologic interest that are well off major drainages and reached only by hiking are listed with the drainages serving the most likely approach routes to them. Each point of interest has a number, and its location and access routes are shown on the Points of Geologic Interest maps (see Plate 7A–D).

WEST SIDE APPROACHES

SKAGIT RIVER DRAINAGE

NORTH CASCADE HIGHWAY (STATE ROUTE 20) WEST OF WASHINGTON PASS

1

CONCRETE

Limestone quarry and Ice Age dam

The town of Concrete was established in 1909 upon the merger of several earlier communities that manufactured cement. The large concrete silos at the west end of town are the most visible relics of this dusty enterprise, which shut down around 1968. The cement workers quarried limestone from large fossiliferous deposits just east of the Baker River dam, at the south end of Lake Shannon. The limestone deposits, part of the Chilliwack River terrane, contain numerous fragments of large fossilized crinoid stems (Figure 38), indicating an age of about 330 million years (Carboniferous). Crinoids, or sea lilies, evolved about 490 million years ago, and numerous species exist today in all the world's oceans. They belong to the sea urchin family, but, following a free-swimming larval stage, attach themselves to the ocean bottom and grow into a plantlike animal.

Figure 38. A crinoid and crinoid fossils in a limestone block from the Chilliwack River terrane (drawing of the block after a photograph in Danner, 1970).

Just east of the Baker River, on the north side of Highway 20, high cliffs of glacial gravels are the remnants of a gravel dam that, during the Ice Age, blocked the Skagit drainage completely. Glacial geologists say that this mass of outwash gravel formed as a tongue of ice from the Cordilleran Ice Sheet advanced eastward up the Skagit drainage from the Puget Lowland and blocked the Skagit River. Outwash gravels from the Cordilleran ice in the Puget Lowland began to fill the lake which

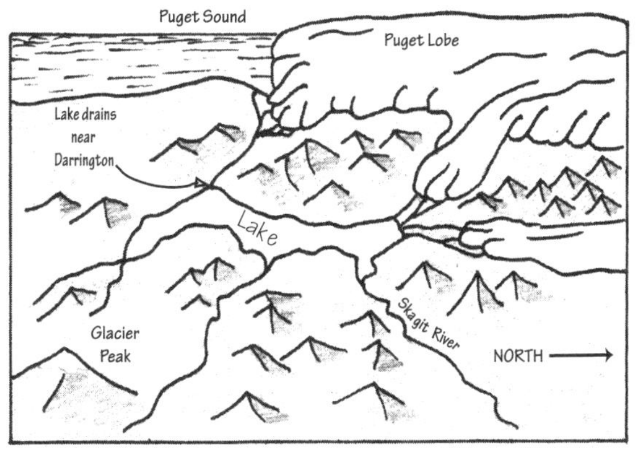

A.

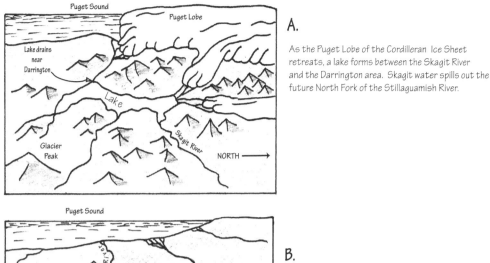

As the Puget Lobe of the Cordilleran Ice Sheet retreats, a lake forms between the Skagit River and the Darrington area. Skagit water spills out the future North Fork of the Stillaguamish River.

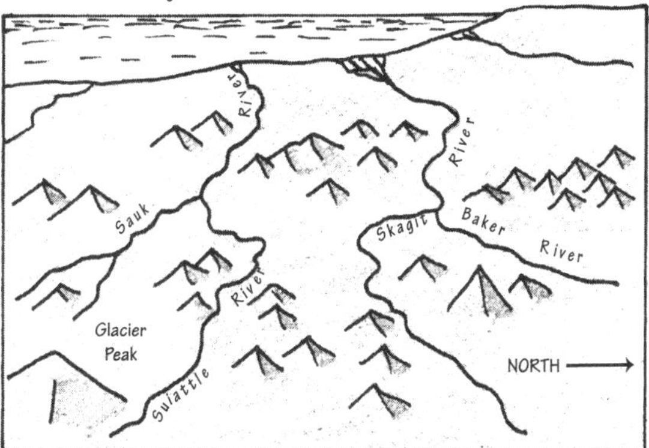

B.

Soon after ice sheet retreats, the Skagit River breaches the outwash plug of gravel and finds its old channel. Sauk and Suiattle Rivers continue to drain out the future channel of the Stillaguamish until the Sauk River builds an alluvial fan which diverts them north, as seen in C.

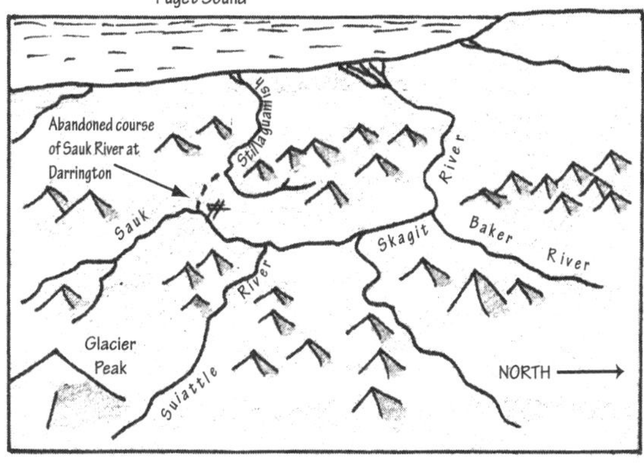

C.

Today both the Sauk and the Suiattle Rivers drain into the Skagit.

Figure 39. How the Sauk and Suiattle Rivers changed their courses.

formed behind the dam. The rising lake waters, still fed by melting ice upvalley, backed up the Sauk and Suiattle Rivers and eventually spilled out into what is now the North Fork of the Stillaguamish River near Darrington. When the Cordilleran Ice Sheet retreated from the Puget Lowland, the lake, followed by the Skagit River, eventually broke through the gravel dam (Figure 39). As deduced by Joseph A. Vance (professor at the University of Washington and a relentless devotee of North Cascade geology, as well as a peak climber), the Sauk and Suiattle Rivers continued to drain through the southern route for awhile. Eruptions of the Glacier Peak volcano at their headwaters eventually so choked the Sauk and Suiattle Rivers with volcanic debris that near Darrington they built an alluvial fan that diverted both rivers north to join the Skagit River.

2

MARBLEMOUNT

Straight Creek Fault

The village of Marblemount sits on a broad valley of glacial outwash. According to a local newspaper, *The Concrete*

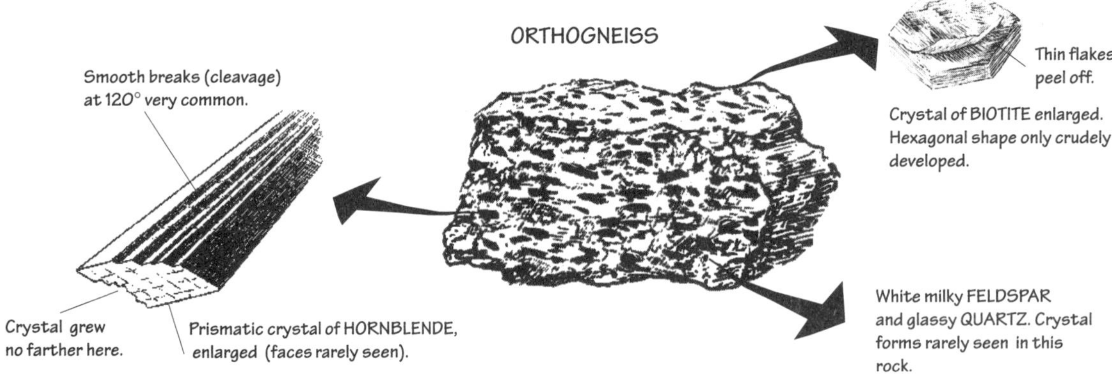
ORTHOGNEISS
Smooth breaks (cleavage) at 120° very common.
Thin flakes peel off.
Crystal of BIOTITE enlarged. Hexagonal shape only crudely developed.
Crystal grew no farther here.
Prismatic crystal of HORNBLENDE, enlarged (faces rarely seen).
White milky FELDSPAR and glassy QUARTZ. Crystal forms rarely seen in this rock.

Figure 40. Minerals visible in tonalite orthogneiss.

Herald, Marblemount is named for an early landmark, Marble Mountain, no longer named on maps (although Marble Creek, about 6 miles to the east, survives). Since the late 1800s, prospectors and other travelers have found supplies and lodging in Marblemount.

Not far east of Marblemount, the outwash gravels hide the double trace of the Straight Creek Fault (Chapter 5), and the traveler looking east sees the rocks of the Cascade Metamorphic Core on Lookout Mountain, rising steep and forbidding. Chopped into the hard metamorphic rock of the core is a maze of zig-zagging logging roads. To the west, on the opposite side of the Straight Creek Fault, are timbered mountains carved from Shuksan Greenschist of the Western Domain (see Plate 2). Ice of the Skagit glacier supplemented by ice from the Cascade River carved the broad valley here at the confluence of the Cascade and Skagit Rivers.

3 SKAGIT RIVER ABOVE MARBLEMOUNT

Major landslide into the river

Upriver from Marblemount, where Highway 20 winds along close to the fast-flowing Skagit, the river is lined with large blocks of rock, many of which interrupt the smooth flow of the river and provide entertainment for whitewater boaters. Travelers here are passing through the remnants of large landslides. If they look closely, they will see in the roadcuts that blocks are jumbled. Several large landslides from ridges to the west and from Lookout Mountain to the east choke the valley here. It is easy to imagine that these landslide deposits once blocked the Skagit River, forming a lake. The idea is confirmed by beds of water-laid volcanic ash in lake sediments a little way upstream of the landslide (8.6 miles from Marblemount; and 5.2 miles from the turn-off to the North Cascades National Park Visitor Center in Newhalem to the northeast). Careful chemical analysis of the ash shows it to be from the cataclysmic eruption of Mount Mazama (Crater Lake) in Oregon some 7,700 years ago. Radiocarbon ages from tree stumps buried in the lake sediments indicate that the lake existed for at least 1,500 years.

4 NEWHALEM

Dating ancient magma using the zircon radiometric clock

The buildings of the town of Newhalem cluster in an orderly fashion on well-groomed lawns across the flat floor of the Skagit River valley. This fastidious landscape contrasts with the wild mountainsides that abruptly tower some 3,000 to 4,000 feet above it. Across the Skagit River, which bounds the town on the southeast, is the North Cascades National Park Visitor Center. Rocks holding up the cliffs surrounding the town and visitor center are metamorphosed igneous plutons (orthogneiss) in the Skagit Gneiss Complex (Chapter 4).

For a good close look at the orthogneiss, leave the grassy park area under the power lines, west of the highway, and walk to the nearest clean granitic cliff. The rock is rough, and its crystals are visibly faceted. With the unaided eye in good light the petrophile can see glassy quartz, milky feldspar with flat faces, black, shiny biotite flakes, and prismatic greenish-black hornblende (Figure 40). The dark minerals are somewhat aligned, like fish in the current of a stream. This orthogneiss, made from a tonalite by metamorphic squeezing and recrystallization (Chapter 4; Geologic Notes 7, 33), is typical of many metamorphosed plutons in the Skagit Gneiss Complex.

Geologists actually know the approximate age of the orthogneiss here, thanks to modern technology and the hard work of geochronologists. In order to determine the age of the rock, the geochronologist grinds up 20 to 100 pounds of rock and, using various methods, some similar to panning for gold, separates out tiny but relatively heavy zircon crystals (Figure 41) from the other minerals. When large and pure, zircon crystals are used in jewelry, but most crystals—common in many granitic rocks—are so tiny that only a flea, or maybe an ant, would value them for decoration. Zircon (zirconium silicate) is a hard mineral and is particularly useful in dating rocks because it resists breakage, melting, and recrystallization during metamorphism. Its real clockwork virtue is that the radioactive element uranium tends to follow the zirconium into the crystal lattice and most zircon crystals thus contain a little uranium. Some atoms (isotopes) of uranium are unstable and gradually decay radioactively at a known rate, changing into various isotopes of lead. After a zircon crystallizes from a magma, lead isotopes from the decaying uranium are trapped in the crystal. By measuring the relative amounts of uranium and lead isotopes in the crystal, a geochronologist can calculate how long ago the crystal formed. The zircon radiometric clock from the orthogneiss at Newhalem has been running about 70 million years. That is to say, the rock first crystallized from a magma in the Cretaceous.

For a distant view of this orthogneiss carved into one of the most rugged areas of the North Cascades, go back (about 0.5 miles) to the bridge over Goodell Creek and look to the northwest to the Southern Picket Range. For a walk through the tonalite orthogneiss in a garden of waterfalls, cross the Skagit River on the bridge to the Seattle City Light Power House and follow the signs to Ladder Creek Falls.

5

THE ANOMALOUS SKAGIT GORGE

Turning a river around

Leaving Newhalem, the highway begins a steeper, winding climb up the Skagit Gorge. The white rocks along the road are mostly tonalite orthogneiss; some are migmatites (Geologic Notes 11, 20). Above Gorge Lake, where the road cuts through banded gneisses rich in layers of mica schist, outcrops are browner and darker.

The gorge of the Skagit contrasts strongly with the broad glacial valley at Newhalem. Geologists have explained this anomalous topography in several ways, but the scenario most consistent with the general pattern in the North Cascades is that the upper Skagit River, including its major tributaries such as Thunder Creek, Big Beaver Creek, and Stetattle Creek, once drained northward into Canada. The gorge is eroded where once there was a bedrock divide (Figure 42). The growth and retreat of successive Cordilleran ice flows brought on this reversal. Each time the ice advanced south into the North Cascades, it dammed north-flowing rivers, forming lakes. Water from the lakes eventually spilled over divides to the south and found new routes to the ocean. When the ice retreated, the lakes reappeared, and some of the sediments deposited in the lakes remain in the valleys to the east. As the ice melted back, lakes that rose high enough to find outlets drained to the south, their rushing waters eroding deep gorges in the bedrock divides. The upper Skagit Lake did just that in the vicinity of Skagit Gorge. Eventually, the new canyon was so deep that even after the Cordilleran ice retreated, the river continued flowing to the south (Geologic Notes 139, 149).

At certain times in the year, travelers will be startled to see little or no water in the gorge above Newhalem. At such times, the entire Skagit River bypasses the gorge by traveling through tunnels from Gorge Dam above to the powerhouse at Newhalem, where turbines turn the steepness of the riverbed into electricity that lights, heats, and cools Seattle.

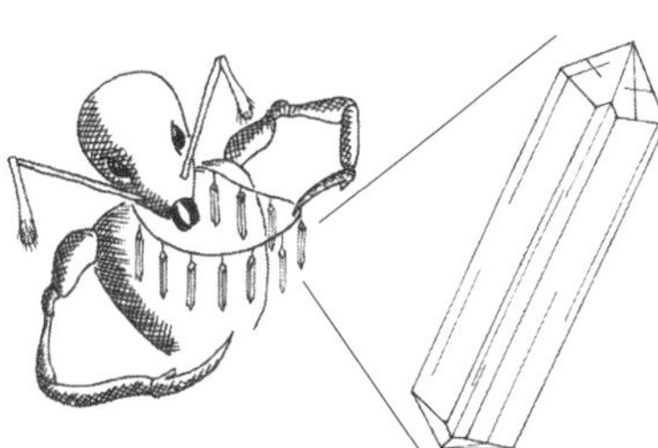

Figure 41. Zircon crystal (greatly enlarged).

6

GORGE CREEK FALLS

Gorge Creek left behind

A visitor stopping to look into the spectacular rock defile of Gorge Creek—after walking out on the equally impressive grated highway bridge!—will appreciate the effects on scenery of the drainage diversion described in Geologic Note 5.

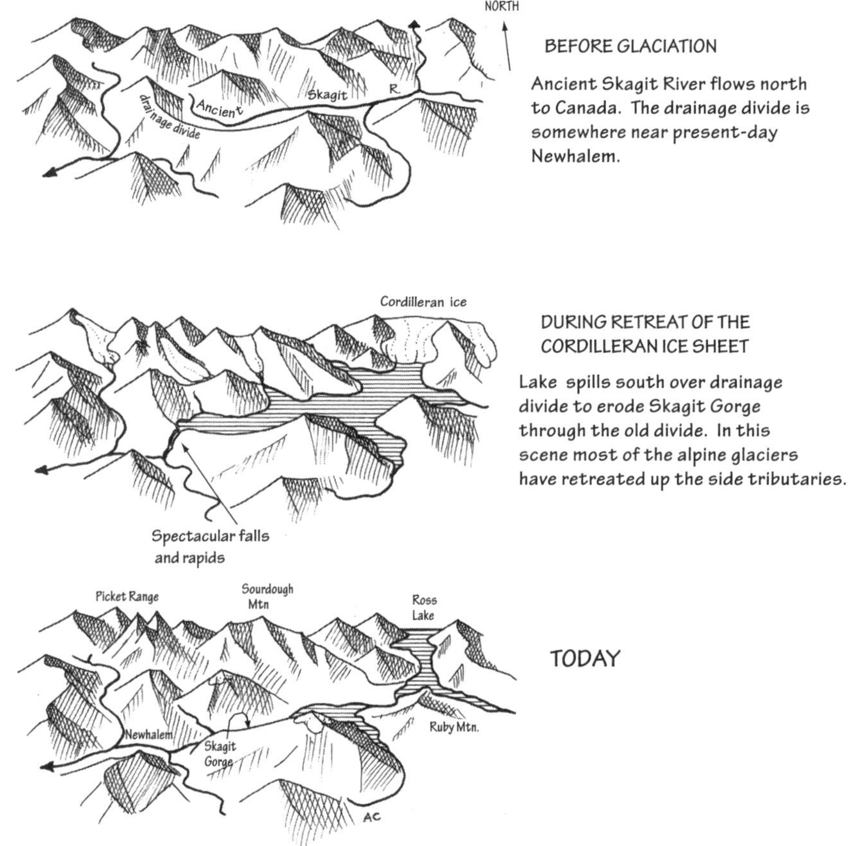

NORTH
BEFORE GLACIATION
Ancient Skagit River flows north to Canada. The drainage divide is somewhere near present-day Newhalem.
Ancient
Skagit
R.
drainage divide
Cordilleran ice
DURING RETREAT OF THE CORDILLERAN ICE SHEET
Lake spills south over drainage divide to erode Skagit Gorge through the old divide. In this scene most of the alpine glaciers have retreated up the side tributaries.
Spectacular falls and rapids
Picket Range
Sourdough Mtn
Ross Lake
TODAY
Ruby Mtn.
Newhalem
Skagit Gorge
AC

Figure 42. Reversal of the drainage of the Skagit River.

Here, a small side tributary, which at one time was gently graded to the main river, hurtles precipitously down to the canyon bottom in a spectacular erosional slot. If visitors could see the headwaters of the creek, they would find a more gentle mountain stream in a small valley. As the Skagit River, augmented by glacially diverted water from the north, rapidly cut a deep gorge down through its former headwater divide, smaller, less powerful tributaries such as Gorge Creek could not keep up and were left hanging above the main gorge (Figure 42).

7

DIABLO DAM

Damming a canyon of stretched granite

Take the turn-off to Diablo Dam for a look at dikes and sills in the Skagit Gneiss Complex, an opportunity to see lineated granite, and a view of the railroad-car lift (elevator) that facilitated construction of Ross Dam. Roadcuts approaching the dam, as well as cuts along the old railroad bed to the top of the railroad lift on the north abutment of the dam, reveal criss-crossing patterns of white ribbons cutting the darker rocks (Figure 43). The first good look at these rocks is on the right (east) side of the road as it winds along the cliffs at the southern abutment. Unfortunately, this is not a good place to stop. The ribbons are really *dikes* and *sills*. A dike is a sheet of igneous rock that filled cracks cutting across existing bedding or metamorphic foliation. If the invading sheet of igneous rock filled cracks running along the same plane as the bedding or foliation, it is called a *sill*. The broad rock slabs of the Diablo Dam spillway also show dikes and sills.

To see an example of rock lineation in the blasted outcrop at the north abutment of the dam, find a conveniently sized piece of orthogneiss, recognized by its black and white minerals and streaky look. Compare the sample to Figure 44, which illustrates the formation of foliation and lineation in rocks. The granitic rocks here are lineated; that is, they have been stretched in one direction only, elongating the minerals into parallel lines when seen from the sides. To be sure the rock is lineated

Figure 43. Sills (vertical white ribbons paralleling rock foliation) and dikes (even whiter criss-crossing ribbons) in outcrop on the road to Diablo Dam.

and not foliated, the observer has to see the structure from three sides, which is not always as easily done with an outcrop as it is with a loose block. Much of the igneous rock that became lineated orthogneiss intruded about 45 million years ago, 15 to 45 million years later than the foliated orthogneisses within the Skagit Gneiss Complex, such as the one described at Newhalem (Geologic Note 4).

To see the railroad lift, walk west along the spur road to an overlook above the town of Diablo. The lift that ascends the cliffs here was first used to lift men and materials during the construction of Diablo Dam, which was completed in 1929. Then, beginning in 1937, supplies for construction of Ross Dam were brought by train to the foot of Diablo Dam. From there, the laden railroad cars were brought up on the lift to the upper level, loaded on barges, and taken up Diablo Lake to the new dam site. A tour of the dams offered by Seattle City Light includes a ride on the railroad lift. Check in Newhalem or Diablo, or call Seattle City Light in Seattle, for information.

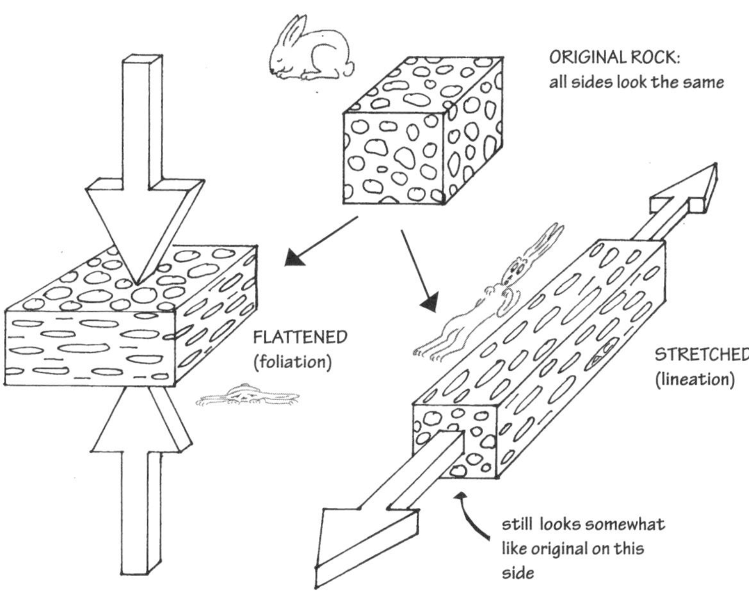

Figure 44. Flattened rock with foliation versus stretched rock with lineation. Combinations of flattening and stretching are more common and more confusing.

THUNDER LAKE FAULT

Broken rock revealed by erosion

After the turn-off to Diablo Dam, Highway 20 ascends a small side canyon containing Thunder Lake (about 23.3 miles from Marblemount). A glacier eroded this side canyon along the shattered rocks of the Thunder Lake Fault. This is one of several north-trending faults cutting the metamorphic and plutonic rocks of the Metamorphic Core Domain. Because the rocks along these faults are shattered and broken, we know that the breaks occurred after metamorphism was long over, when the rocks were cold and could no longer recrystallize. Commonly, faults reveal themselves to geologists by juxtaposing rocks that generally do not belong next to each other, or because a bed or unique rock marker is cut off by the fault and reappears somewhere on the other side of the fault-trace (Chapter 5; Figure 25). No such markers are found along the Thunder Lake Fault, which is only revealed by the shattered rocks along the fault and the long, narrow erosional valleys following its trace (Geologic Notes 2, 12, 13, 26, 44, 72).

COLONIAL CREEK CAMPGROUND

Mudflows and exotic blue-green waters

Rhode Creek is a small tributary that is building an alluvial fan into the drowned Thunder Creek valley. Colonial Creek Campground sprawls across this fan. Alluvial fans are mostly built by catastrophic mudflows that roar down side-stream canyons during unusually heavy rainfall or rapid snowmelt. In the 1980s a mudflow descended Rhode Creek and covered the highway near the campground. A new layer of gravel and boulders was added to the fan. Most of this new deposit has been removed from the highway, but some fresh bouldery remnants along the side remind us of geologic process.

A more constant sedimentation process is filling Diablo Lake at the end of Thunder Arm, where Thunder Creek is rapidly building a delta into the lake. The sandbars and islands of deposited sand and gravel can be seen from the bridge, but better views of this aggregation can be found along the Thunder Creek trail.

Another, but much slower, silting of Diablo Lake takes place as very fine glacial silt carried by Thunder Creek settles across the entire lake bottom. This fine silt suspended in the water causes the unusual milky turquoise color of the lake. Boaters on Diablo Lake approaching Ross Dam will be struck by the abrupt change in water color when they cross from the water dominated by Thunder Creek to water from the upper Skagit River coming out of Ross Dam powerhouse. Ross Lake is free of the glacial silt color, except from very near west-side tributaries that drain valleys with glaciers. Most streams that feed into the lake have ice-free headwaters.

DIABLO LAKE OVERLOOK

Rocks and glaciers behind the scene

From the parking area at the Diablo Lake Overlook (1.5 miles along Highway 20, traveling east from the bridge over Thunder Arm of Diablo Lake), look southwest to views of glaciers on Colonial, Snowfield, and Pyramid Peaks (Figure 52). These mountains are carved from banded gneiss in the Skagit Gneiss Complex, also described in the display at this overlook. Look to the west across Diablo Lake to see the long, straight-but-small valley of Sourdough Creek, climbing steeply up to Sourdough Mountain. This erosional trench is carved along the crushed rock of the Thunder Lake Fault (Geologic Note 8).

Prior to the Pleistocene Ice Age, the upper Skagit River and major tributaries such as Thunder Creek drained north to the Fraser River in Canada. During the Ice Age, Cordilleran ice blocked the river's northward course, and its dammed waters spilled south to erode the Skagit Gorge (Chapter 7; Geologic Note 5). The level benches and terraces prominent east of Diablo Resort and where you now stand were probably part of the old valley bottom when the waters all ran to the north. The deep gorge of upper Diablo Lake was eroded by the voluminous meltwater coming off the Cordilleran ice as it advanced and retreated. During the peak of glaciation the valley was filled with ice over a mile thick, which also rasped away at the rock and rounded many of the landforms.

HORSETAIL CREEK

The mixing of migmatite

After crossing the bridge over Horsetail Creek (Pierce Falls), about 1.7 miles beyond the Diablo Lake Overlook, stop on the shoulder to view classic outcrops of migmatite. Here are light- and dark-colored gneisses in complex layered masses, cut by thick and thin white dikes, lacing

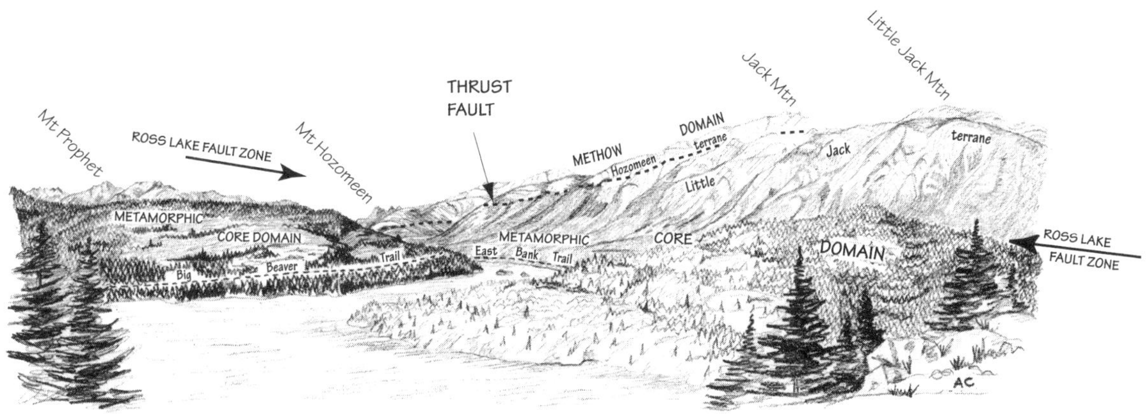

Figure 45. View to the north from the Ross Lake Overlook on Highway 20. The main strand of the Ross Lake Fault Zone is under Ross Lake near and parallel to the East Bank Trail.

the rock in all directions. The gneisses may all be highly deformed—squeezed—igneous dikes or sills (Geologic Note 7), but some could have been sandstone beds before they were thoroughly metamorphosed. Migmatite means mixed rock, and in the North Cascades migmatite usually consists of a mixture of light- and dark-colored metamorphic and igneous rocks in such a complex interlacing of different types that describing them is not easy. Here, migmatites appear to have been made by magma invading mica schists derived from sandstone and shale. Dikes and sills invaded the rock, then the entire mass was moved plastically and swirled. This was followed by more dikes, perhaps followed by more movement, then more dikes and so forth (Geologic Notes 20, 116).

Professor Donna Whitney has determined that all of this took place at great pressure induced by depths of 18–19 miles (30 km) in the Earth's crust, probably at temperatures as high as 1,330 F° (720 C°). Walk south and around towards Pierce Creek to outcrops of hornblende schist and mica schist, which are remnants of the material invaded by numerous magmas. The schist can be recognized by its dark colors and abundance of the minerals hornblende and/or biotite. Also to be found are small pods of dark slippery chlorite rock or serpentinite, derived from the Earth's mantle (Geologic Note 67).

12

WESTERN ROSS LAKE OVERLOOK

Views of faulted terranes

Ross Lake is a hydroelectric reservoir behind Ross Dam, mainly constructed between 1937 and 1949. The lake is 22 miles long and ends just beyond the United States border at the 49th parallel (see Plate 2). From the Western Ross Lake Overlook (about 3.3 miles east of the Diablo Lake Overlook), look north up Ross Lake (hopefully on a clear day) to view the Ross Lake Fault (Chapter 3), which separates the Metamorphic Core Domain of the North Cascades from the Methow Domain to the east (Figure 45). On the west side of the lake, high ridges carved from the Skagit Gneiss Complex lead up to Mount Prophet (7,660 feet). On the east, Jack Mountain (9,066 feet), carved from basalt of the Hozomeen terrane, towers some 7,400 feet above the lake. The lower slopes of Jack Mountain are underlain by mica schist and other rocks of the enigmatic Little Jack terrane (Chapter 2), which is separated from the overlying Hozomeen terrane by a thrust fault.

The metamorphic minerals in the Skagit Gneiss Complex west of the Ross Lake Fault formed at much hotter temperatures (about 720° C) than those in the Hozomeen greenstones (about 300° C) east of the fault. This differential indicates that the Skagit Gneiss has been uplifted from much deeper in the crust than the Hozomeen rocks.

For other stories of fault uplift, see Geologic Notes 41 and 65.

13

GRANITE CREEK

The mystery of its straight valley

Geologists are suspicious of any long, straight topographic feature, commonly suspecting erosion along the crushed rocks of a fault to be the cause of the straightness. Such breaks in the rock tend to be straight for several, if not hundreds, of miles. So it is with the valley of Granite Creek, which Highway 20 follows for several miles. (The highway makes a pronounced bend to the south about 6.3 miles east of the Western Ross Lake Overlook. It continues south to the end of the valley at Rainy Pass). For somewhat random erosional forces to produce a straight valley, they must be acting on pronounced and persistent features in the bedrock. The crushed rock of a fault zone erodes more easily than the surrounding rock; hence the fault itself is worn away faster than rocks to either side (Plate 4C; Geologic Notes 2, 8, 26, 44). A contrast in rock types across such a straight valley would be even more convincing evidence that the feature is a fault, but the rocks on either side of this valley are both granite of the Golden Horn batholith. The valley parallels some major faults in the region, and the granite is locally crushed and altered along the highway. A significant fault would offset the contacts of the Golden Horn batholith, yet no offset has been observed (see Plate 2). This valley may be eroded out along a fault, but geologists are unsure of the postulated fault's significance in the overall scheme of things.

14

CRATER MOUNTAIN

Views of an ancient ocean floor

Travelers heading northwest on Highway 20 where it follows Granite Creek view the great mass of Crater Mountain rising 6,000 feet above the valley bottom of Canyon Creek. To enjoy the same view, however, east-bound travelers should stop at the viewpoint and sign located 11.6 miles from the Western Ross Lake Overlook; or 11.0 miles from Rainy Pass, and look back the way they came (Figure 46). Crater Mountain is carved from greenstone of the Hozomeen terrane, once the basaltic floor of the ancient Methow Ocean (Chapter 3). The mountain's name probably derives from the crater-like shape of its summit, which was carved from the ancient metamorphosed basalt by modern cirque glaciers. The rugged battlements of the greenstone viewed from the highway rest above phyllite of the Little Jack terrane, which is exposed in the lower slopes of the mountain. The thrust fault that separates the two terranes here was one of the first such faults discovered in the North Cascades by the late Peter Misch (Chapter 2).

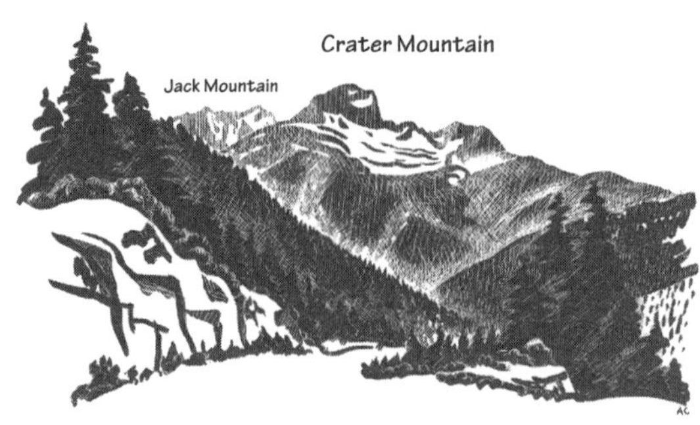

Figure 46. View of Crater Mountain, looking northwest from Highway 20.

15

BLACK PEAK BATHOLITH

Wandering of a geologic compass

East of Rainy Pass, the highway descends to a low point on a curve above Bridge Creek and swings by outcrops of broken greenish-gray granitic rock of the Black Peak batholith. This old pluton holds up many rugged peaks such as Whistler Mountain, directly north, and, naturally, Black Peak to the west. Travelers who stop to examine the rock will notice that it looks sick. Its minerals are not crisp white and black, as in many old plutons (Geologic Note 4), but shades of green, and fuzzy. The rock has been altered to water-rich minerals like chlorite. A visitor might notice also that the rock is full of bore holes about one inch in diameter. These are not artifacts of highway building (such drill holes are about 2 inches in diameter), but the tracks of "paleomagicians" (Chapter 5). The Black Peak batholith (Geologic Note 46) has been scrutinized by geophysicists trying to decide if it is part of a terrane that traveled north great distances, after it crystallized some 90 million years ago (late Cretaceous).

In order to determine the paleomagnetism of a rock such as this—that is, to determine the orientation of the Earth's magnetic field at the time when the rock was formed—the paleomagnetist first collects a sample of it, carefully keeping track of the sample's orientation. The easiest way to secure such a sample is to cut a core from the rock using a diamond drill powered by a chainsaw motor and measure the precise orientation of the core relative to the north pole and the horizontal. The paleomagnetist then takes the oriented

Figure 47. View to the south from the Washington Pass Overlook at peaks carved mostly from granite of the Golden Horn batholith.

core back to the laboratory and carefully determines the rock's magnetic field. Commonly, the rock's magnetism is complex—as if a magnetic tape had been used repeatedly, yet with incomplete erasure of the previous recording—and the paleomagnetist will, bit by bit, demagnetize the sample, measuring its magnetic field at each stage of demagnetization. The Black Peak batholith has been continually magnetized by a changing field for the last 90 million years, but fortunately, the magnetization acquired in its formative years, so to speak, is the most resistant to the demagnetization. By stripping off the younger magnetic influences, the paleomagnetist can reveal the rock's original magnetic orientation.

Plutons in the North Cascades similar to the Black Peak batholith yield paleomagnetic directions that suggest they crystallized thousands of miles to the south of where they are today. Unfortunately, the Black Peak batholith has proved to be a poor example. Although early work on the Black Peak batholith suggested it too had traveled long distances, later analysis of the data indicated that its paleomagnetism could not be reliably measured.

Some geologists dispute this scenario of peripatetic plutons, arguing that the batholiths formed more or less where they are and have since been tilted. The paleomagnetic inclination of the rock can reveal the latitude of its birth (see "Paleomagnetism: Finding a Rock's Place of Birth" in Chapter 5), but if the pluton has been later tilted, the measured inclination can be a reflection in part or whole of the tilting. In bedded rocks, such as lava flows, for instance, the original horizontal position can be determined. Not so with a uniform granitic batholith. The doubters argue that even when the magnetism is well measured—as with other batholiths, not the Black Peak—the possibility of tilting renders the original position on the Earth uncertain.

16

WASHINGTON PASS OVERLOOK

Granite inspires

At Washington Pass, turn north into the visitor center parking lot and walk a short distance out to the overlook for spectacular views of peaks and spires carved from granite of the Golden Horn batholith (Figure 47). The distinctive pinnacled ridges owe their look to joints in the rock. Joints are geologist jargon for cracks, plain and simple. But the processes that produce cracks are many. Granitic batholiths cool and crystallize under considerable pressure within the Earth. Upon reaching the surface, through uplift and erosion, they expand, causing many cracks to form, sometimes in a pattern reflecting the shape of the batholith. Joints in granitic rocks are commonly at right angles, and where vertical joints predominate, weathering produces serrated peaks and ridges. Some of the more prominent notches, such as those separating Liberty Bell, Early Winter Spires, and other pinnacles, are not joints, but small faults, as indicated by broken and ground-up granite along them. The blocks of rock on either side have moved at least a little, and erosion has worn away the broken rock. The same faults may be seen cutting the granite in the towers of Kangaroo Ridge and the Wine Spires on the north ridge of Silver Star Mountain, to the east.

For more information on the Golden Horn granite, see Geologic Note 138.

ROADS AND TRAILS REACHED FROM THE SKAGIT RIVER AND BEYOND

17

SAUK MOUNTAIN

Views and remnants of old volcanoes

For a rewarding side trip east of Concrete, take the Sauk Mountain Road (6.5 miles east of the Baker River bridge and 0.6 miles west of the entrance to Rockport State Park). The road climbs seemingly endless switchbacks (but actually only 7.9 miles) up a very large landslide, eventually reaching a parking area and trailhead for Sauk Mountain. From this spot, hang-glider pilots launch themselves for a wonderful ride down to fields along Highway 20, 4,000 feet below. The views from the parking lot are spectacular. For even better views, climb the switchback trail to the summit of Sauk Mountain ($^3/_4$ mile; 1,000-foot elevation gain). The trail climbs up through cliffs of volcanic rocks of the Chilliwack River terrane. Along the lower switchbacks the metamorphosed volcanic rock is highly foliated, but along the summit ridge, look for volcanic breccia of mostly green, gray, and black angular fragments of volcanic rocks surrounded by dark green or gray matrix (Figure 48). This breccia was deposited in a submarine fan at the toe of a volcano in the ancient Chilliwack volcanic arc, perhaps 250 million years ago (Permian).

From the summit, look south across the broad valley floor at the confluence of the Skagit and Sauk Rivers. Here is a good place to contemplate the lake that filled these valleys (Geologic Note 1) when the Skagit drainage was dammed by glacial debris at Concrete. The lake drained via the North Fork of the Stillaguamish near Darrington.

To the east, much of the mountainous interior of the North Cascades National Park and Ross Lake National Recreation Area is in view on a clear day, as are rocks of all three geologic domains of the North Cascades (Figure 49).

18

BERDEEN LAKE

An ancient ocean and the Cascade Arc

Berdeen Lake lies in a glacier-carved cirque. At one time the lake may have been much larger, but the upper end has been filled in with gravels, much of it outwash from the glaciers on Hagan Peak (Figure 50). The lake, perched high above a lower cirque on the west fork of Bacon Creek, illustrates beautifully the double-cirque or stepped-valley pattern of the North Cascades and other temperate mountain ranges (Chapter 7).

Figure 48. Volcanic breccia of the Chilliwack River terrane on the summit ridge of Sauk Mountain.

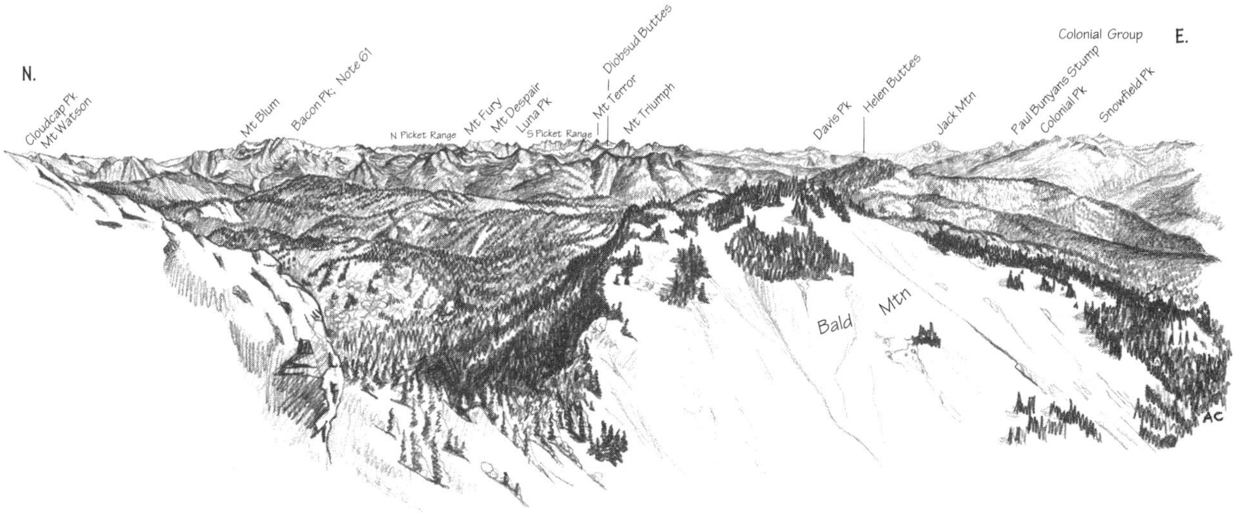

Figure 49. Looking northeast from Sauk Mountain at high peaks, mostly in North Cascades National Park. All three domains of the North Cascades are visible, but the faults bounding the domains go from left to right in this drawing and are mostly hidden behind the ridges. Bald Mountain, Cloudcap Peak, Mount Watson, Mount Blum, Bacon Peak, Diobsud Buttes, and Helen Buttes are in the Western Domain. All the other named peaks and ranges are in the Metamorphic Core Domain, with the exception of Jack Mountain, which is in the Methow Domain.

At the upper end of Berdeen Lake, the gravelly valley floor conceals the contact between Darrington Phyllite—black, layered with white quartz veins—and massive light-colored, slightly pinkish granodiorite of the Mount Blum pluton, a Cascade Arc root pluton of the Index family (Chapter 6). The contact is mimicked by the debris in the till, which consists of dark phyllite from Hagan Peak and light-colored granodiorite from ridges and peaks to the north.

The muds which became Darrington Phyllite were deposited on the ocean floor some 160 million years ago (Jurassic). One hundred and thirty million years later, the granodiorite invaded the phyllite. The ongoing movement of Pacific Ocean plates against the North American continent recreates this collage of rocks over and over as deposits in the ocean are swept under the continent's edge or plastered on it by subduction and then eventually intruded by melted rock generated by the same subduction process.

Berdeen Lake is seldom visited, and to do so requires a rigorous climbing backpack over trailless terrain, from either the logging road in Bacon Creek or the Baker River via Blum Lakes. For a route description, see Beckey, 1995.

19

THORNTON LAKES

On the edge of the batholith

On reaching the high ridge (elevation 5,050 feet above lower Thornton Lake; about 4 miles and 2,500 feet of climbing beyond the Thornton Lakes trailhead), the hiker may want to pause and view a principal geologic contact in the cirques beyond. The hard steep climb up from the trailhead passes through small promontories and cliffs of granodiorite, typical outcrops of granitic terrane in the western North Cascades. Looking into the Thornton Lake cirque, the discerning viewer sees dark amphibolite (metamorphosed basalt), with some layers of orthogneiss forming a cliff on the right (east). At the bottom of the cliff and facing toward the viewer, highly fractured dark reddish to brown outcrops are mica quartz schist (metamorphosed chert). The amphibolite and metachert are metamorphosed oceanic rocks of the Napeequa Schist (Chelan Mountains terrane). These dark metamorphic rocks contrast strikingly with light-colored granodiorite of Mount Despair on the slopes west (left) of the lake. The granodiorite of Mount Despair is one of the largest plutons of the Chilliwack

Figure 50. Berdeen Lake. The contact between dark phyllite and light-colored granite is hidden under the glacier, outwash, and till at the head of the lake. The till on the glacier reflects the change in rock types under it.

batholith. It invaded the Napeequa Schist about 32 million years ago (Oligocene), not long after the Cascade Volcanic Arc began building volcanoes (Chapter 6). The contact between these rocks lies under the water of Thornton Lakes, but can also be viewed near Thornton Creek, above the lake. Mount Triumph, made of old orthogneiss, forms an impressive backdrop.

20

SOURDOUGH MOUNTAIN TRAIL

Views of Skagit Gneiss

Where the highway crosses the bridge over Gorge Lake, it passes the turn-off to the company town of Diablo. Behind the town, the Sourdough Mountain Trail climbs high above the glaciated valleys for unparalleled views of rugged peaks carved from rocks of the Skagit Gneiss Complex. The hiker who has reached the crest of Sourdough Mountain (about 5,000 feet of climbing up about 5.2 miles of steep trail) can view many rocks of the Metamorphic Core Domain (see Plate 2). Close up are outcrops of migmatitic banded gneiss to delight the geologic heart. Brown layers of biotite schist alternate with white sills of tonalite orthogneiss and lineated granite (Geologic Note 7), as well as pegmatite—made of very coarsely crystalline feldspar and quartz (Figure 51; Geologic Note 116). The schist layers, remnants of the Chelan Mountains terrane, began as shale on the ocean

Figure 51. Migmatite of mica schist (dark) and granitic sills (light).

bottom, but locally pieces of mantle (ultramafic rock) from beneath the ocean floor have also been incorporated into the terrane. North of the lookout, pods of emerald green metamorphosed ultramafic rock can be seen (Geologic Note 67). Metamorphism and multiple injection by igneous sills at great depth in the crust have produced the complex seen here.

The views from the summit ridge of Sourdough Mountain (Figure 52), especially at the site of the fire-lookout, have inspired many, including poet Gary Snyder:

> "*...The little cabin—one room—*
> *walled in glass*
> *Meadows and snowfields, hundreds of peaks.*"
>
> Gary Snyder, *The Back Country*

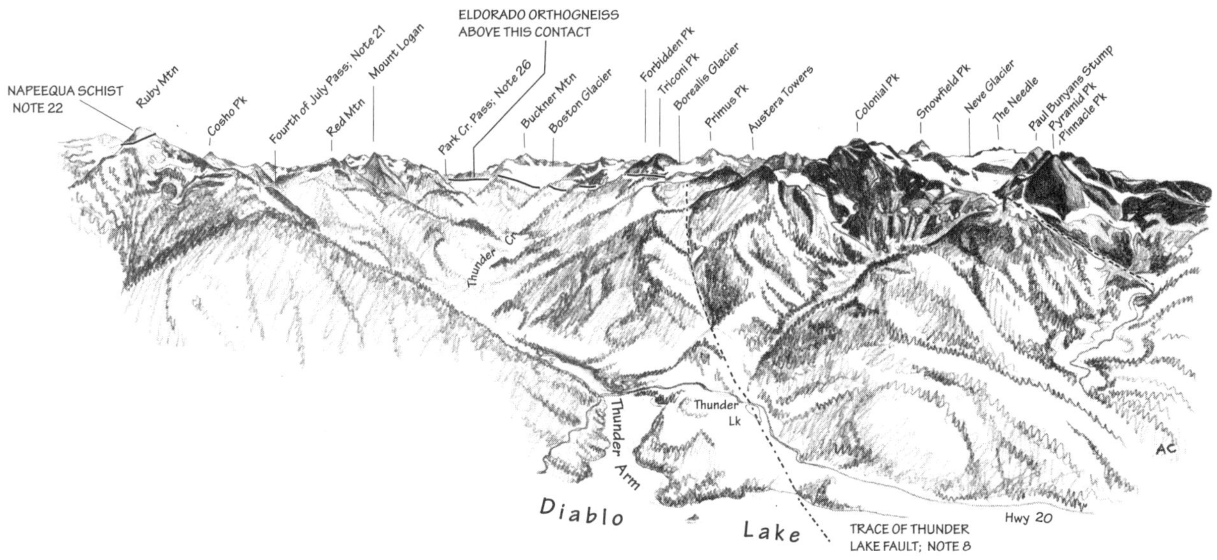

Figure 52. View to the south from the Sourdough Mountain Lookout. Except where noted, the scene is mostly eroded from the Skagit Gneiss Complex.

THUNDER CREEK TRAIL AND BEYOND

21

FOURTH OF JULY PASS

Rocks far from home

There is not much to see in the broad, tree-covered notch of Fourth of July Pass (5 miles and a 2,300-foot climb from Colonial Creek Campground), but the trail from the west climbs across interesting rocky benches carved from the Skagit Gneiss Complex and crosses several terraces mantled with river gravels perched high above today's Thunder Creek. The terraces are probably remnants of the preglacial river that flowed north to the Fraser River (Geologic Note 5). The pass is a glacier-cut notch in the dominant ridge that forms Ruby Mountain and Red Mountain. Ruby Mountain often stood high above a sea of Pleistocene ice as a *nunatak,* or rock island, in the ice. In the woods are boulders of conglomerate, rich in dark chert pebbles. These *erratics*—glacier-carried boulders of a rock type foreign to the place where they are found—came from the Methow Domain, many miles to the northeast.

22

RUBY MOUNTAIN

A long climb and a huge fold

The summit of Ruby Mountain is a marvelous viewpoint for North Cascades National Park, but the crude steep trail (about 2.5 miles from Fourth of July Pass) and 4,000-foot climb entice few visitors. Rocks on the summit are fine-grained amphibolite (metamorphosed basalt) and mica quartz schist (metamorphosed chert), as well as pods of ultramafic rock (metamorphosed mantle), commonly colored red-orange by weathering. This is the typical assemblage of oceanic rocks (Napeequa Schist) of the Chelan Mountains terrane (Chapter 3), which are here perched atop gneiss and schist of the Skagit Gneiss Complex. Beyond Ruby Mountain to the northeast is the Ross Lake Fault, and beyond it is relatively unmetamorphosed rock of the Little Jack and Hozomeen terranes. Geologists and others too may find it useful to think of the metamorphic core of the North Cascades as a huge, albeit complex, fold (Figure 53) that is faulted on the sides. Almost a mile below the top of Ruby Mountain, once deeply buried Skagit migmatites are exposed along Highway 20. The piece of the Napeequa Schist on Ruby Mountain is a remnant of the core's less-metamorphosed husk, which is preserved here up on the northeastern side of the huge fold.

23

THUNDER CREEK TRAIL

Evolution of the Thunder Creek scene

In the reaches above the delta at Thunder Arm, Thunder Creek flows through steep-walled chasms, whereas hikers travel on relatively level benches, commonly far above the creek. Thunder Creek truly does thunder, creating a din more to be attributed to a river than to a creek. Thunder Creek valley and its tributaries reflect a history of glacial scour, drainage diversion, and changing base levels (Chapter 7). The trail on the benches and terraces follows the bottom of the glacier-carved valley, whereas the river has sawed a slot below the valley floor, probably in response to a change in its base level, which occurred when the upper Skagit River drainage was reversed to the west (Geologic Note 5).

24

CONFLUENCE OF THUNDER AND FISHER CREEKS

The Great Dismal Swamp

At Fisher Creek, the trail climbs an alluvial fan and glacial debris to gain a ridge between Fisher and Thunder Creeks. The trail gradually ascends the ridge leaving Thunder Creek to meander placidly through swampy meadows almost 2,000 feet below. Why this contrast between the rock gorges below the Fisher Creek junction and the gentle gradient of Thunder Creek above? Probably there are several factors, but the most important is the alluvial fan built by Fisher Creek where it empties into Thunder Creek. The fan essentially dammed Thunder Creek, allowing sediment to accumulate upvalley. In the early history of this dam, the broad swampy meadows were probably a lake (Geologic Note 29).

25

FISHER CREEK

Landslide from Ragged Ridge

The Fisher Creek Trail crosses hummocky, bouldery ground about 7.5 miles from Easy Pass, the site of a

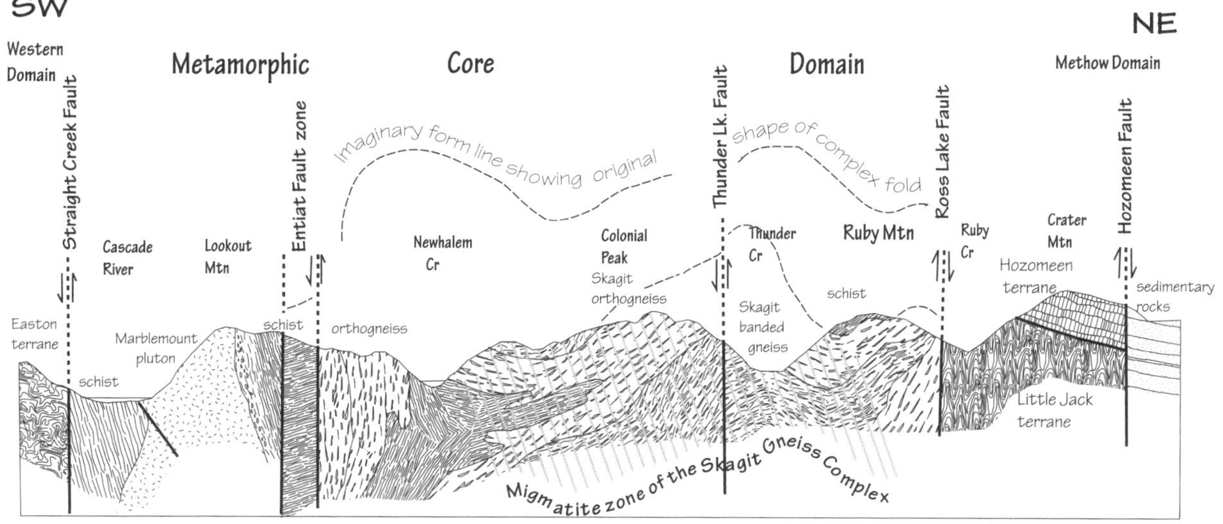

Figure 53. Cross section of giant fold exposing migmatites in the Metamorphic Core Domain (Chelan Mountains terrane). The flanks of the fold are faulted.

monster rockslide that came down from Ragged Ridge. Near an old, ramshackle trapper's cabin, next to a very large boulder, look out over Fisher Creek to brushy rock-strewn desolation and appreciate the magnitude of the slide. Scientists think that many such monster slides roared into the valleys during major earthquakes (Geologic Note 73).

26

PARK CREEK PASS

Faults and prospects

Park Creek Pass is an impressive alpine defile eroded along an old fault. The trail climbs to the east of the trenchlike pass to avoid permanent snow. The rocks in the defile are fault-smashed tonalite gneiss. Even though these rocks appear hard and durable they are more easily eroded than the unfaulted gneiss forming the high peaks and surrounding summits. Peculiar ridges and slots across the ridge west of the pass indicate similar parallel faults. Prospectors have burrowed into the fault zones in the high meadows above Park Creek, especially west of the pass. The sulfide minerals that they found were probably left in the rock by hot waters escaping through the fault-shattered rock from crystallizing magma of the Cascade Pass dike (Geologic Notes 64, 127).

BIG BEAVER TRAIL

27

ROSS LAKE BOAT LANDING

Finding marble in the Skagit Gneiss

Much of the Skagit Gneiss Complex that holds up the mountains at the south end of Ross Lake is so thoroughly metamorphosed, and has been so thoroughly injected by magma, that even the experienced geologist cannot always be sure how the rocks began (Geologic Note 11). Blasted roadcuts at the edge of Ross Lake, next to the Ross Lake Resort boat dock, reveal gray marble, a metamorphic rock derived from limestone. Although the marble is lighter-colored on broken surfaces, the dark layers reveal the layering of the original sedimentary beds. Most limestone begins life in a body of water at the Earth's surface, probably as accumulated shells (made of calcium carbonate) of marine animals (Chapter 2). We think that the schists of this outcrop were also derived from sedimentary rocks, but there are also plenty of white granitic dikes cross-cutting the marble and schist. How can the lighter-colored marble be distinguished from the light-colored igneous dikes? The marble scratches readily with a knife or ice ax; the feldspar and quartz of the dikes do not.

28

PIERCE CREEK

Orthogneiss outcrops

Creeks coming down the northeast side of Sourdough Ridge have sharply incised the lower valley walls, and the trail from Ross Dam to Big Beaver Creek crosses these erosional saw cuts. Climb down to the outcrop from the trail crossing at Pierce Creek to find nice clean orthogneiss (Geologic Note 4) of the Skagit Gneiss Complex. The tonalite orthogneiss here is made of white feldspar and quartz, with conspicuous crystals of black biotite and hornblende. This orthogneiss can be seen along the Big Beaver Trail in blocks fallen from the hillside and rare outcrops in the first mile or so beyond Ross Lake.

29

BIG BEAVER SWAMPS

A hanging valley dammed

The Big Beaver Trail wanders along the north side of the Big Beaver Creek valley, staying mostly on dry alluvial fans or mossy rocks of the valley side in order to avoid the soggy ground of the Big Beaver swamps. The swamps and ponds are all but hidden from the trail by vegetation. Naturally, some big beavers have contributed to the sogginess, but geologic factors are the prime cause. Big Beaver Creek flows in a typical glacier-carved, U-shaped valley (Chapter 7). Prior to the filling of Ross Lake reservoir, Big Beaver Creek flowed in a hanging valley not far above the main Skagit River valley (Geologic Notes 31, 105), spilling in falls or rapids over a bedrock lip into the valley below. It still does, but only just, because Ross Lake has all but drowned the rapids. Hikers who look downvalley from the Big Beaver bridge can see the remnants of the rocky rapids in the river just a few feet above lake level.

Side streams debouching onto the flat valley floor of Big Beaver Creek built fans, which tended to dam the main creek, at least temporarily. The creek, like the trail, was forced to the north side of the valley by a large fan on the south side near Ross Lake. The bedrock at the mouth of the creek and ongoing accumulation of sediments behind the alluvial dams maintains the especially flat valley floor, good for swamps and beavers.

LITTLE BEAVER CREEK TRAIL AND BEYOND

30

LOWER LITTLE BEAVER CREEK VALLEY

Deep-ocean deposits along the trail

The Little Beaver Creek Trail switchbacks steeply above Ross Lake to gain the hanging floor of Little Beaver Creek valley (Geologic Note 31). Where the trail begins to level off, hugging the precipitous mountainside above the gorge of Little Beaver Creek, outcrops of chert of the Hozomeen terrane (Chapter 3) provide an opportunity to rest. Born in the deep ocean some 220 million years ago (Triassic), this radiolarian chert forms from an accumulation of quartz skeletons of microscopic marine plankton. A good hand lens will reveal the tiny rounded forms of the radiolarians in some of the siliceous rocks. Wetting the rocks helps bring out the tiny blebs. The rock is layered with thin shaley beds between thicker quartz-rich layers. These rocks formed in an ocean deep far from any land, where the arrival of silt or mud eroded from either continents or volcanic islands is a rare event that is possible only through the most violent, far-reaching density flow (Chapter 3). The rest of the time only the remains of marine animals sprinkle down from the ocean above as generations live and die.

A paleontologist liberates the very small shells from this submarine cemetery by slowly dissolving the rock with dilute acid. Once identified, the fossils may reveal the age of the deposit and tell us something about the environment of the ancient ocean.

31

UPPER LITTLE BEAVER CREEK VALLEY

Hanging valleys

Little Beaver Creek valley is a splendid example of a glacier-carved, U-shaped valley. During the last alpine ice advance in the Pleistocene Epoch (beginning some 25,000 years ago), a glacier filled the valley to the top of the vertical cliffs (Figure 54). Smaller glaciers in valleys tributary to Little Beaver Creek could not cut as deep as the main glacier. As a result, the bottoms of the tributary valleys (such as those of Perry, Redoubt, and Pass Creeks) remain hanging high above Little Beaver Creek. And Little Beaver Creek itself hangs high above the Skagit valley (Geologic Note 30).

32

MIDDLE LAKES

Bog iron

A deposit of reddish bog iron is continually being formed along the eastern shore of the southeast Middle Lake. Ground water passing through the granodiorite west of Tiny Lake leaches iron out of small crystals of *pyrite* (iron sulfide) and deposits the iron as limonite at springs near the lake.

33

MOUNT REDOUBT

Banded gneiss

Two features are conspicuous in the Skagit Gneiss Complex exposed in Bear Lake basin and on Mount Redoubt: the striking bands, which are layers containing different amounts of light- and dark-colored minerals, and foliation, the platy, streaky look of the rocks caused by parallel alignment of their minerals, especially mica. (For information on how to reach this area check Tabor and Crowder, 1968, and Beckey, 1995). The banding in the gneiss may in part be bedding, inherited from the original sedimentary and volcanic rocks (Chapter 4). The dark layers are rich in the minerals hornblende and plagioclase and may have been basalts. The lighter layers contain more biotite and quartz and may have been sandstone. Many light-colored layers are intrusive sills of tonalite or granodiorite, now metamorphosed to orthogneiss.

The parallel alignment of minerals is a characteristic of metamorphic rocks that were deformed as their crystals grew. Crystals grow in one direction more readily than in others; flakes of biotite grow at the edges to make larger flakes, and needles of hornblende grow at the end to make longer needles. Under the great heat and pressure of metamorphism, rocks behave more like taffy than hard stone. Elongated crystals rotate as the rock flows around them, often coming to rest with the long dimensions of the crystal aligned with the flow direction. Some crystals may break along minute parallel cracks, allowing the tiny pieces of crystals to rotate until they too lie parallel to the direction of flow in the rock. These processes also align the light and dark layers in the rock so that eventually everything becomes aligned or almost aligned, like pages in a book (foliation) or the pasta in a package of uncooked spaghetti (lineation). Imagine mushing out a mixture of pennies (flat crystals, like mica) and nails (elongate crystals, like hornblende) in taffy (flowing rock). Messy, but aligned. Slippage in some layers in the Mount Redoubt area was so intense that most crystals were broken down, and only very small crystals survive. Such rocks are usually dark colored and streaky, and sometimes contain rounded white crystals of feldspar that were strong enough to survive the deformation process (Figure 55).

Figure 54. In the Pleistocene, a valley glacier filled Little Beaver Creek to the top of the lower cliffs.

34

NORTHERN PICKET RANGE

Veins and joints make the landscape

Climbers on the Picket Traverse, traveling the high slopes above Picket Creek, see how rock features influence the mountain scene. (For information on how to reach this area, see Tabor and Crowder, 1968, and Beckey, 1995.) Planes of weakness and joints in the biotite orthogneiss of the Crooked Thumb–Phantom Peak ridge influence the shape and form of the ridge eroded from the rock (Figure 56). Some of the vertical joints are filled with veins of quartz and small amounts of sulfide minerals containing copper (yellow chalcopyrite), iron (silver-yellow pyrite), lead (silver-colored cubes of galena), and arsenic (dull silver arsenopyrite).

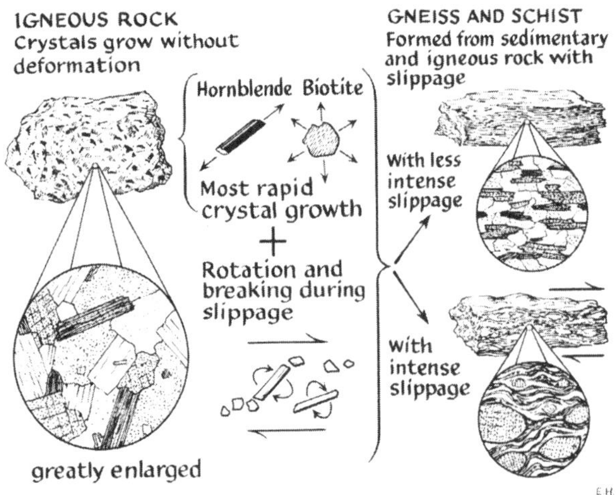

Figure 55. Foliation forms as crystals grow aligned during deformation.

The gneiss overlies a large mass of once-molten tonalite, an arc-root pluton of the Chilliwack batholith. Vapors and water that carried the metals "boiled off" the molten mass and percolated through vertical joints in the overlying gneiss roof. As the emanations cooled and/or reacted chemically with the roof rock, the minerals crystallized in the veins.

EAST BANK TRAIL (ROSS LAKE) AND BEYOND

35

HIDDEN HAND PASS

Ancient valley bottom

Hikers traveling the trail along the east side of Ross Lake will not be cheered by the 900-foot climb over Hidden Hand Pass, although the rise to the high point of the pass is gentle. The pass itself is hardly noticeable in the gently sloping forests. The broad shoulder surmounted by the trail may be one more terrace remnant of the earlier north-flowing Skagit River, a survivor from the time before the new channel to the south allowed the river and its tributaries to cut the present valleys below Hidden Hand Pass (Geologic Note 5). It could also be a drainage channel cut when a glacier filled the Skagit valley and water ran along the side of the ice long enough to erode the valley walls. Or a combination of both.

36

LIGHTNING CREEK

A gorge from the past

The high steel bridge over the steep-walled inlet of Lightning Creek emphasizes the drainage changes that have so stamped the landscape in the Ross Lake area. Most of the side streams tributary to Ross Lake have cut deep slots as their local base level was lowered when the upper Skagit River was diverted to flow into Puget Sound (Geologic Note 5). Many slots are now flooded by the reservoir (for example, Devils Creek; Geologic Note 105), but Lightning Creek has an even more complicated history (Geologic Note 37).

37

DESOLATION PEAK

Ancient oceanic rocks and a view of stream piracy

The hiker ascending Desolation Peak climbs through rubbly rocks of the Hozomeen terrane—marine basalt, chert, and shale. Fossils found in chert and rare marble beds indicate that the rocks here are about 250–300 million years old (Carboniferous and Permian). Much of the valley holding Ross Lake has been carved from rocks of the Hozomeen terrane (Chapter 3).

Sixty sunsets had I seen revolve on that perpendicular hill. The vision of the freedom of eternity was mine forever. The chipmunk ran into the rocks and a

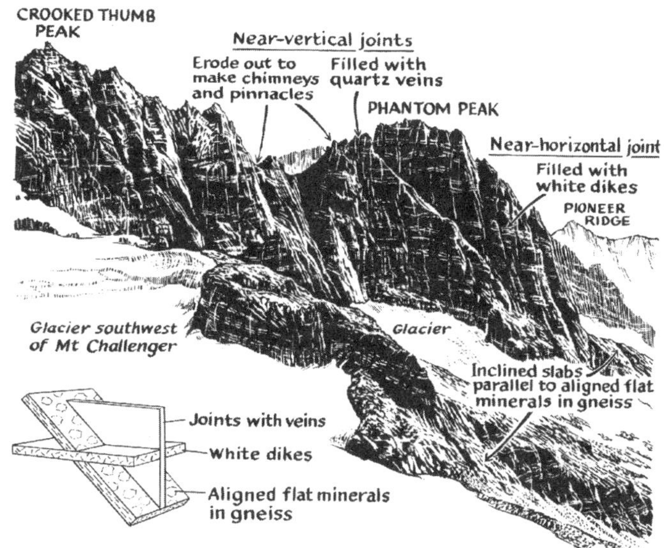

Figure 56. Joints, dikes, and foliation control erosion and influence the scene.

butterfly came out. It was as simple as that.

So wrote Jack Kerouac in his autobiographical novel, *The Dharma Bums,* in which the narrator, seeking spiritual enlightenment, spends a summer as a fire lookout on Desolation Peak.

From the top of Desolation Peak (4.5 miles and 4,500 feet above Ross Lake), others seeking enlightenment have a good view of the great chasm of Lightning Creek to the east. Lightning Creek has grown by a series of stream diversions that have considerably changed

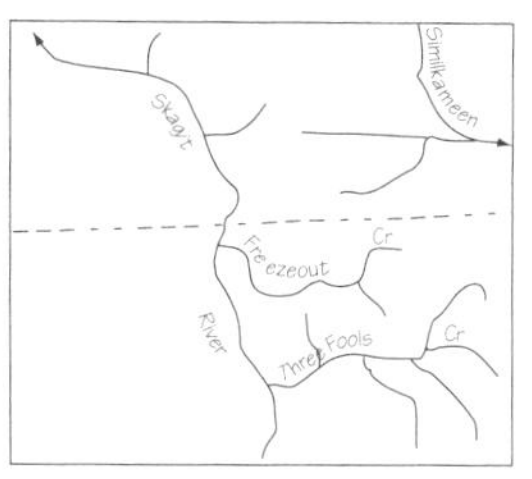

Prior to advance of Cordilleran ice, Skagit River drains north. Tributary drainage is established.

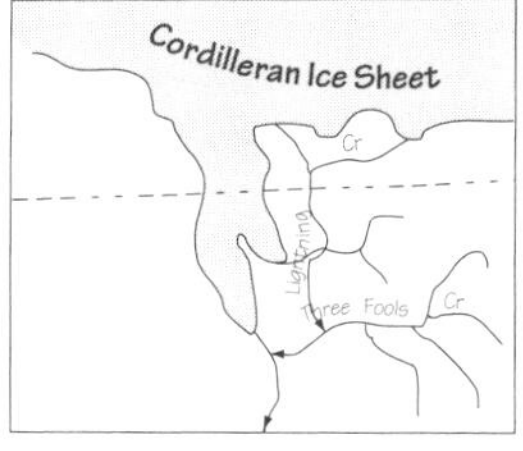

As Cordilleran Ice Sheet retreats, meltwaters on the east cut new channel across established drainage divides. Lightning Creek is established.

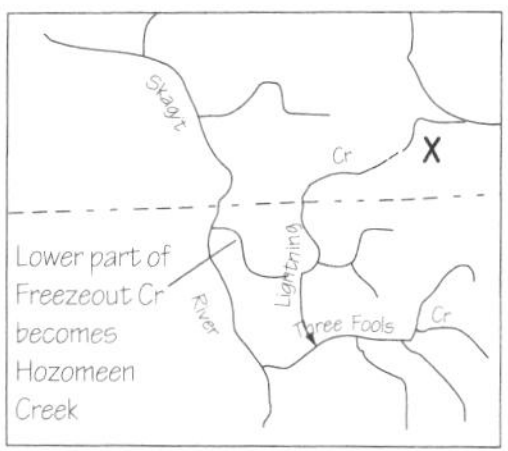

Lightning Creek continues to extend northward at **X** by capture of Similkameen tributaries.

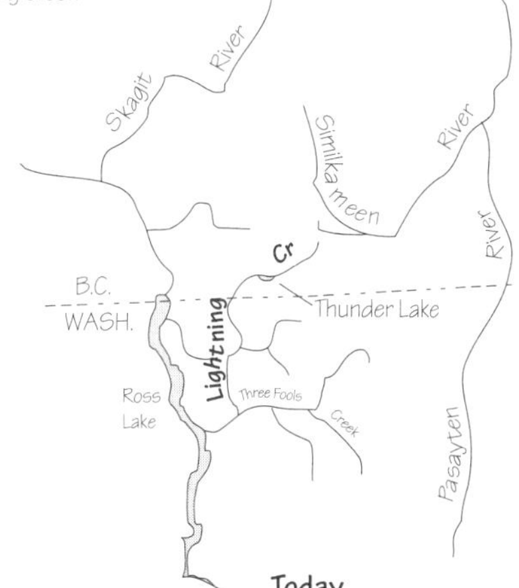

Figure 57. Lightning Creek, born of glacier meltwater, may yet capture more tributaries of the Similkameen River.

the drainage of the area (Figure 57). At one time, prior to reversal of the Skagit drainage by the Cordilleran Ice Sheet (Geologic Note 5), Freezeout Creek drained west through the pass at Willow Lake and out Hozomeen Creek to the Skagit River (what is now Ross Lake reservoir). The upper reaches of Lightning Creek in British Columbia drained northeast into the Similkameen River.

When the Cordilleran Ice Sheet was retreating, meltwater in the vicinity of the Similkameen, northeast of what is now Ross Lake, spilled south down a tributary to Freezeout Creek. Freezeout Creek itself was probably blocked by Cordilleran ice lingering in the deep Skagit valley. A lake may have formed and spilled south into Three Fools Creek. The abundant meltwater cut down easily in the sandstone and argillite of the Methow Ocean rocks. The drainages that had been flowing across the hard greenstones of the Hozomeen terrane were left high, if not dry. Even after the ice was all gone, the new and deep gorge of Lightning Creek was established. Freezeout Creek was beheaded, its lower course becoming Hozomeen Creek. The northward extension of Lightning Creek by headward erosion continued to capture water from the Similkameen drainage. Little by little, the piratical Lightning Creek reversed the Similkameen tributary. Lightning Creek's voracious progress has been temporarily arrested by a series of rockfalls that have dammed the waters to form Lightning and Thunder Lakes in British Columbia, but in time more Similkameen water will flow south.

38

DEVILS DOME

Glacial-polished conglomerate

Hikers on the trail over Devils Dome (about 7.0 miles from Ross Lake and 4,400 feet of climbing above the lake) may detour to the ridgecrest north of Bear Skull Shelter to find great outcrops of glacier-polished coarse conglomerate. Look for rounded boulders of light-colored granitic rock in the conglomerate. These boulders were derived from granitic uplands to the east and carried out into the deeper water of the Methow Ocean by density flows (Geologic Notes 144 and 150). The slurries of mud and sand that carried the cobbles rushed out onto the ocean floor about 105 million years ago (Cretaceous).

39

LITTLE JACK TRAIL

Ascending mysterious rocks for a good view

The Little Jack Trail was built by the Forest Service in the early part of the twentieth century in order to graze pack animals in the meadows on Little Jack Mountain. The trail zig-zags up a brushy hillside, occasionally crossing outcrops of granitic rock and fine-grained mica schist. Little is known about these granitic rocks except that they occur as a swarm of small igneous intrusions in a belt of schist making up the Little Jack terrane (Chapter 3). Hikers who reach the meadows and panoramic views near the summit of Little Jack Mountain will find outcrops of mica schist, some cut by white or orange dikes of fine-grained volcanic rock. The summit ridge of Little Jack offers a good view north to the towering cliffs of Jack Mountain, which are mostly made of greenstone. These metamorphosed oceanic basalts of the Hozomeen terrane formed the floor of the Methow Ocean. They have been thrust-faulted over the Little Jack terrane. The fault crosses the ridge in the low notch at the foot of the Jack Mountain escarpment (Geologic Notes 12, 14).

McMILLAN PARK AND JACKITA RIDGE TRAILS AND BEYOND

40

CRATER MOUNTAIN TRAIL

Pillow basalt of the Hozomeen terrane

One guidebook writer says the western and true summit of Crater Mountain possesses one of the best views in the North Cascades, well worth 6 miles of switchbacked trail and the 6,200-foot ascent, with a final rock scramble. On the other hand, another team of guide writers opine that nothing could make the hike worthwhile, but they know nothing of rocks. Even if the difficult western summit is not obtained (or the lower and less-steep eastern summit is the goal), the hike up through cliffs of greenstone is geologically interesting. Along the trail above treeline, look for rounded, pillowy forms in the greenstone; that is, metamorphosed basalt (Figure 58). These *pillow basalts,* as they are called, offer definitive evidence that the lava forming the basalt erupted under water—in this case, under the ocean. The pillows form by a budding process.

Figure 58. Pillow basalt (greenstone) of the Hozomeen terrane near the east summit of Crater Mountain. Note characteristic radial joints in broken pillows.

As the molten rock oozes out onto the ocean floor, it chills immediately on its surface but remains molten inside. The molten interior continues to exert pressure on the hardened bubble, which cracks, letting magma ooze out to form a new bubble that hardens in turn. The pillows so formed mostly break off and roll down the slope to collect as rubble in the deep. The globular pillows are commonly mixed with angular blocks of basalt to make volcanic breccia (Geologic Note 17), which is also abundant in the greenstone.

41

McMILLAN PARK TO DEVILS PARK

Hozomeen Fault

The trail between McMillan Park and Devils Park mostly lacks rock excitement, and its aggravating descent into the head of Nickol Creek before climbing up into the meadows of Jackita Ridge might dampen the ardor of any rock-loving hiker. Even so, the topographic low itself is a significant geologic feature. Just east of the lowest point and a small knob, the trail swings into a small Devils Creek tributary which has eroded along the main strand of the Hozomeen Fault (5,320 feet; about 6.3 miles from the trailhead in Canyon Creek). We suspect a fault strand goes through the low saddle as well, but at the eastern strand, the traveler passes from greenstone of the oceanic Hozomeen terrane into sedimentary rocks of the Methow Ocean (Geologic Note 42). In Chapter 3, we noted that these sedimentary rocks were deposited on Hozomeen rocks forming the floor of the Methow Ocean. The ocean floor here has been uplifted many thousands of feet on the Hozomeen Fault (Figure 59). Pods of serpentine crop out along the Hozomeen Fault and some can be found in the creek. Reaching these outcrops requires a short trailless hike

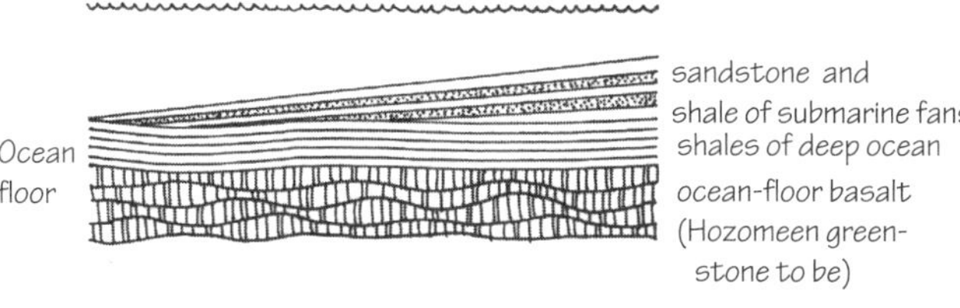

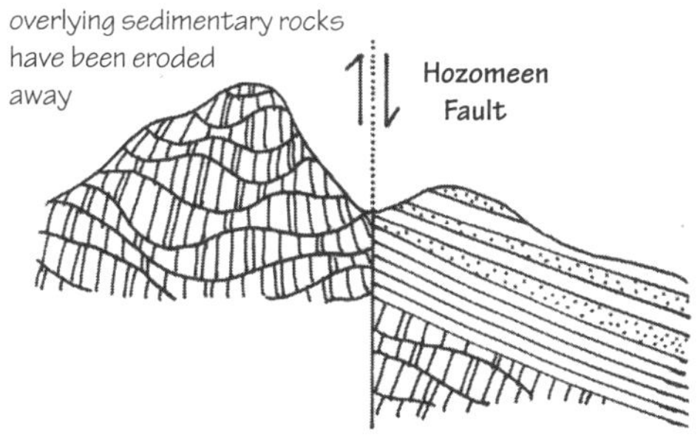

Figure 59. Hozomeen greenstone of the ocean floor faulted up relative to ocean sediments of the Methow Domain.

downstream. Apparently the fault has dragged up some slivers of mantle.

42

JACKITA RIDGE TRAIL

Conglomerate near Anacortes Crossing

Travelers on the Jackita Ridge Trail north of Devils Park and approaching Devils Pass can detour along an abandoned trail to Anacortes Crossing to look at wonderful conglomerate consisting mostly of chert pebbles. The chert pebbles are smooth white, gray, or black, and very hard, with a surface that resembles porcelain. About 90 to 95 million years ago, these cobbles and boulders were eroded off the emerging Hozomeen terrane to the west—the one-time floor of the Methow Ocean—and deposited by streams flowing across the sediment that filled the ocean (Chapter 3).

TRAILS REACHED FROM GRANITE CREEK

43

CANYON CREEK

Gorge and gold

The Canyon Creek Trail climbs high above the deep, steep-sided Canyon Creek gorge. Lowering of the local base level when the upper Skagit River was diverted to the southwest (Geologic Note 5) and ice-age diversion of much of the Pasayten drainage into Canyon Creek by way of Holman Pass (Geologic Note 149) have both contributed to the sawing of this spectacular slot.

The trail winds in and out of side canyons, all eroded from slate and fine-grained sandstone. Sedimentary bedding in these highly folded rocks is only conspicuous here and there, but mainly hikers are likely to be struck by the monotony of these deep ocean deposits. Below, in the gorge, things have been more exciting for prospectors and miners. The sands of Canyon Creek have been the source of considerable placer gold, discovered by itinerant prospectors in the 1870s. At times, hikers may hear the noise of an engine from below in the canyon, where modern miners are sucking gold from the creek's crevices with large power vacuums. The gold has been eroded out of quartz veins in rocks of the Methow Domain upvalley (Geologic Notes 143, 148).

44

EASY PASS

A groovy fault

Easy Pass was eroded where the rocks of Ragged Ridge are crushed along an east-west fault. Rocks on the south side of the pass are speckled granitic rock of the Black Peak batholith (Geologic Notes 15, 46). To the north are metamorphosed siltstone (now dense, fine-grained brownish-purple rock) and conglomerate of the Methow Domain, here intruded by numerous dark dikes. In places these rocks were stretched during metamorphism, and the conglomerate pebbles are now cigar-shaped. A little way to the northeast is light-colored Golden Horn granite (Geologic Notes 16, 138).

To see crushed and grooved rock in the fault zone, leave the trail about a quarter-mile east of the pass and walk across to the south wall of the gully. Be careful of very steep snow along the base of the wall. Not every part of the wall will be grooved; it may take some looking. The grooves were gouged along the fault as the block on one side scraped along the block on the other side. That the grooves are horizontal indicates the fault movement was also horizontal. A comparison of the offset of rock units on either side reveals that the south side of the fault moved eastward relative to the north side.

45

CUTTHROAT PASS

Granite sand

At Cutthroat Pass on the Pacific Crest Trail (about 5 miles and 2,000 feet above the trailhead at Rainy Pass to the west, or 5.5 miles and 2,300 feet from the roadhead below Cutthroat Lake to the east), the views of granite peaks carved from the Golden Horn batholith are grand. Hikers can also look along the trail to discover deposits of granitic sand. This material is not a stream deposit (as is much sand), but an accumulation of granite crystals and clots of crystals broken from the granite by weathering. In exposed areas like this, large temperature fluctuations cause individual minerals to expand and shrink a little, but because each of the different minerals in the rock shrinks or expands a little differently than its neighbor, the changes tend to break the minerals away from each other. In addition, biotite grains in the granite absorb water, and their consequent expansion pries the grains apart. Geologists call the sandy material thus produced *grus*. In some desert

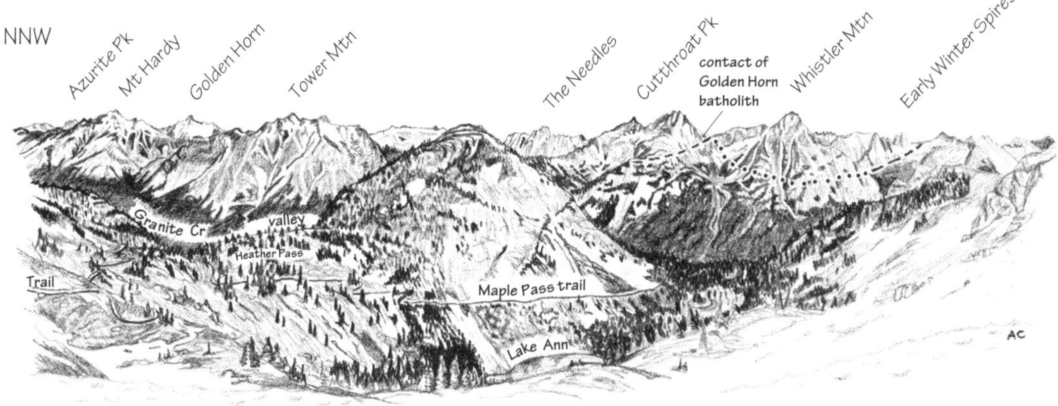

Figure 60. View to the east from above Lake Ann. Mountains on the far side of Granite Creek and beyond the dashed contact line are carved from granite of the Golden Horn batholith. The dotted line represents hidden contact behind Whistler Mountain. Foreground ridges are tonalite of the Black Peak batholith.

climates, the grus developed on granite is tens of feet deep. Hikers who pick up a handful can pick out shiny, flat-sided feldspar crystals, glassy quartz, and flaky black biotite even with the naked eye.

46

MAPLE PASS

Views of a wandering batholith

Hikers who have reached the high point (6,980 feet) between the Lake Ann and Rainy Lake cirques (about 4 miles from the trailhead and about 2,000 feet of climbing, via Maple Pass) not only have terrific views in all directions but can contemplate two batholiths, one a foreign immigrant and one a native Washingtonian. Black Peak and its attendant ridges display dark cliffs and glacier-sculpted ridges, known as arêtes. Orange and yellow peaks rise to the east, across Granite Creek valley and the highway (Figure 60). The dark-hued mountains are eroded from the tonalite and quartz-diorite (Chapter 6; Figures 86 and 112) of the Black Peak batholith; the orange-to-yellow peaks eroded from the granite of the Golden Horn batholith. Black Peak magma invaded the surrounding rocks of the Chelan Mountains terrane about 90 million years ago (Cretaceous), but the Golden Horn granite intruded only about 45 million years ago (Eocene). Some geologists believe that the Chelan Mountains terrane may have been located far to the south when it was invaded by Black Peak magma (Geologic Note 15), but was more or less where it is today by the time Golden Horn magma rose up into it.

WEST SIDE APPROACHES

BAKER RIVER DRAINAGE

BAKER RIVER ROAD (USFS ROAD 11) AND BEYOND

47

BURPEE HILL

The Puget Lobe outwash

Travelers bound for Baker Lake by way of the Burpee Hill Road have opportunity to view thick deposits of well-bedded sand and gravel deposited about 17,000 years ago (Pleistocene), when a tongue of the Cordilleran Ice Sheet advanced up the Skagit valley, and sediment-loaded rivers emerging from the glacier snout deposited sand and gravel in front of the glacier. The road winds in and out of alcoves in these unstable deposits, each alcove the bite of a landslide. A layer of unsorted rock debris and glacial sand and silt (the stuff of ice-borne till) tops the well-bedded sands and gravels and can be seen sometimes in the cut bank at the top of the grade (Figure 61). The outwash, deposited as the glacier advanced, and the till cap show that the outwash was overridden by the advancing ice. Because the roadcuts continually slump and change, the till cap may be hard to find.

48

VIEWS OF MOUNT SHUKSAN

Greenschist and phyllite

Weather permitting, a wonderful view upvalley of Mount Shuksan surprises the traveler ascending the Baker River Road (7.4 miles from Highway 20). Rising with startling abruptness and glistening white above the alder and second-growth fir on the glacial deposits of the Baker River valley, Mount Shuksan seems a different world. And indeed it is. Mount Shuksan is an erosional remnant of a once-continuous layer of greenschist and phyllite making up the Easton terrane, which was thrust over the Chilliwack River terrane, Bell Pass Mélange, and Nooksack terrane. The latter underlies the Baker River valley here. Erosion-resistant Shuksan Greenschist stands high in the crags of Mount Shuksan (Plate 8C; Geologic Notes 79, 80).

49

BAKER RIVER VALLEY

Basalt erupted under the ice

Strange cones of rock rising above the flat floor of the Baker River valley (about 11.4 miles from Highway 20) are erosional remnants of young basaltic deposits. These deposits formed about 90,000 years ago, when hot basalt oozed out under glacial ice in the valley and exploded into a mass of fine, glassy debris, building the large piles. The hot glass of the quenched basalt lava was immediately altered to distinctive low temperature minerals in the watery environment.

50

BAKER RIVER

A river misplaced

For a good view of the Baker River Gorge below Upper Baker Dam, drive across the dam and park at the eastern abutment. Walk back across the dam, completed in 1959, and view the deep gorge, eroded in rocks of the Nooksack terrane. The bedrock gorge is a surprise considering what appears to be a broad gentle valley all around. In fact much of the lower river canyon, now mostly obscured by Lake Shannon behind the Lower

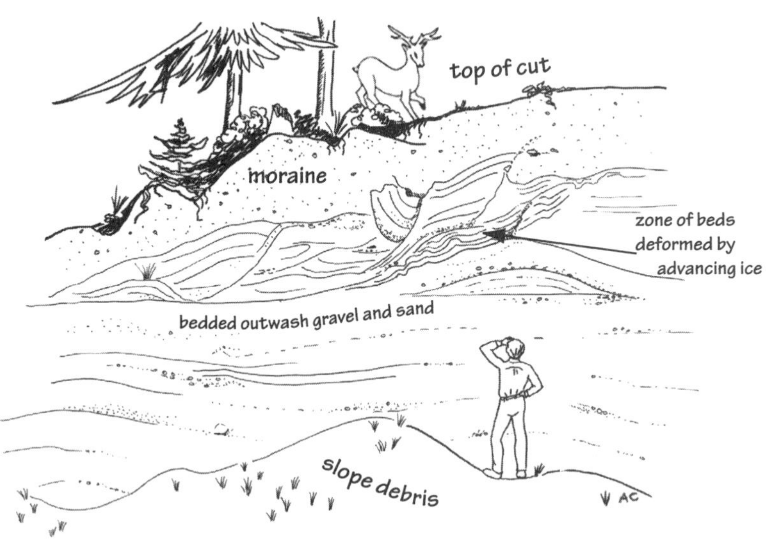

Figure 61. Roadcut on Burpee Hill Road reveals glacial till overlying advance outwash (stream deposits) from Cordilleran glacier.

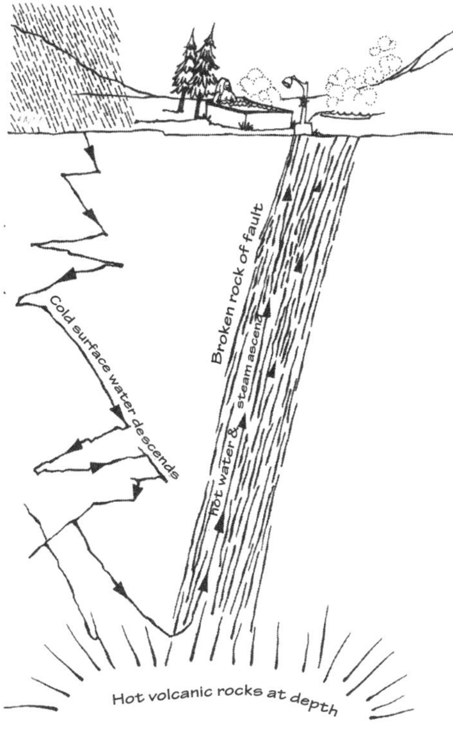

Figure 62. Broken rock of fault zone facilitates circulation of water up from hotter rocks (from Tabor, 1987).

Baker River Dam at Concrete, is cut in glacial outwash. During retreat of the Pleistocene ice in Baker River valley (about 13,000 years ago), a thick deposit of clay, sand, and gravel filled the old valley. When the glaciers withdrew, the river wandered back and forth, as rivers tend to do, on a broad outwash plain (Chapter 7). Sulphur Creek lava erupting from the small cone near Schriebers Meadows (Geologic Notes 53, 55, 59) flowed out onto this outwash plain, forcing the Baker River to the east side of the valley. When the outwash dam near Concrete had been flushed away by the Skagit River, and the river returned to its old route to the ocean (Geologic Note 1), it quickly removed the gravels that filled its valley. The smaller Baker River could not keep up with its more powerful trunk river and was soon left high on its gravels. A series of falls and rapids down to the Skagit worked back up the Baker River. The river down-cut like a saw in wood, making a slot in the outwash plain. The position of the Baker River on the broad, filled valley was independent of its original channel, established before the gravels filled the valley. And like a saw in wood, the river just had to keep cutting into whatever it hit in its slot. At both dams the river was positioned over bedrock ridges, not over its original channel—a bit of luck for the dam builders, who like to attach their structures to bedrock.

51

BAKER HOT SPRING

Tapping Mount Baker's heat

A short side trip from the Baker River Road via USFS Road 1144 (22.2 miles from Highway 20) brings dirty hikers to a parking area for Baker Hot Spring (3.2 miles from USFS Road 11). An unmarked, well-scarred trail leads to the springs. The sulfurous spring testifies to volcanic heat still present in the crust below Mount Baker, but the circulation of water to this heat may be facilitated by a major fault on the west side of the Baker River valley (Figure 62), which separates rocks of the Nooksack terrane from those of the Chilliwack River terrane (see Plate 2). Unfortunately, because past bathing revelers left such a mess, a once-charming large wooden tub has been destroyed, and the spring is now an unimproved shallow pool. Expectant hikers who arrive at the springs will probably leave dirty.

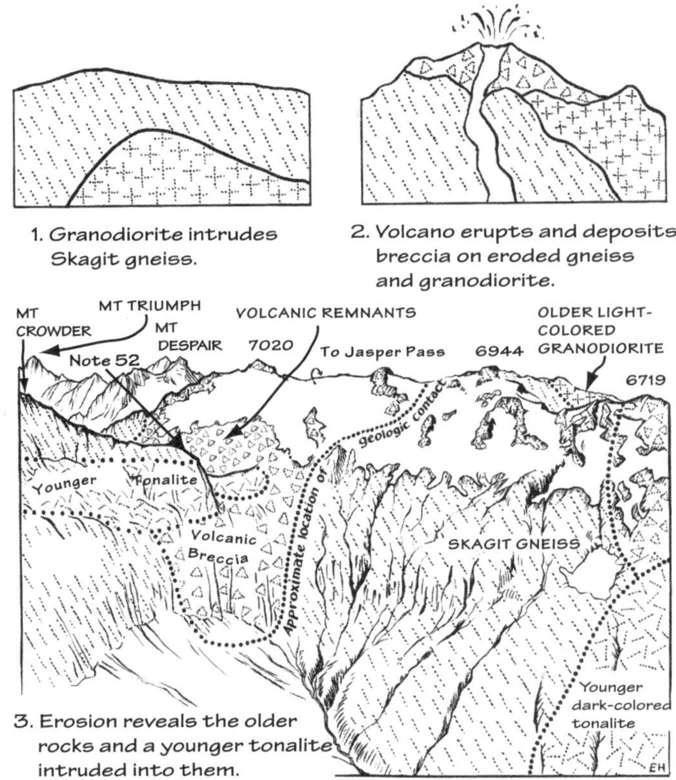

Figure 63. View of Pioneer Ridge from the west. Cross sections show (1) intrusion of granodiorite of the Index family into Skagit Gneiss and (2) deposition of volcanic breccia on top of eroded granodiorite. In today's scene (3) erosion has revealed plutons of younger tonalite of the Snoqualmie family that intrude the older rocks.

52

PIONEER RIDGE

Finding the Cascade Volcanic Arc

Hikers following the Pioneer Ridge High Route (Tabor and Crowder, 1968; Beckey, 1995) in the vicinity of Mount Crowder will have good opportunity to examine the products of the Cascade Volcanic Arc, both the volcanic eruptive materials and the deep residuals of the arc's plumbing system.

A breccia composed of angular fragments of lava, gneiss, and granitic rocks enveloped in a greenish matrix underlies the slopes above Picket Creek. This breccia contrasts strikingly with the surrounding light-colored granodiorite of Pioneer Ridge to the west and the darker-colored tonalite north of the nearby lake. Parts of all these rock types—the breccia, the granodiorite, and the tonalite—were molten when each formed, but the relationships at the contacts between the various rocks, as well as the shapes, types, and relations of minerals in them (as revealed through a microscope), show they formed at different times. The light-colored Index-family granodiorite (Chapter 6) solidified deep in the Earth's crust, was uplifted and eroded into mountains, and then was covered by the breccia, which is a remnant of volcanoes that erupted somewhere in the vicinity some 26–33 million years ago (Oligocene).

Other remnants of the volcano are on Pioneer Ridge, where blocks of the underlying granodiorite in the lowermost layers of the breccia show that the volcano blanketed hills of granodiorite and slopes of talus and scree. About 24 to 25 million years ago, tonalite magma of the Snoqualmie family intruded the breccia (Figure 63). Erosion has revealed all these rocks and sculpted them into today's mountains.

SCHRIEBERS MEADOW ROAD (USFS ROADS 12 AND 13) AND BEYOND

53

SCHRIEBERS MEADOW ROAD

Sulphur Creek lava flow and its future

Three miles beyond the turnoff from the main Baker River Road (USFS Road 11), the road to Schriebers Meadow (USFS Road 12) ascends the sloping surface of a basaltic lava flow, which erupted from a small parasitic volcano on the slopes of Mount Baker at Schriebers Meadow. Rocky and Sulphur Creeks flow on this relatively young lava and have cut steep-sided slots in it. To glimpse the rubbly surface of the flow, continue on USFS Road 12 toward Wanlick Pass to an old burn, about 1 mile from the junction with USFS Road 13. Since the rough surface of the flow has not been glaciated, geologists know that it erupted after the Ice Age ended about 13,000 years ago. Because the lava is so permeable, rain and snowmelt have little chance to erode it before seeping underground. Between major creek channels the lava is well preserved. Unfortunately the brush is thriving and even a glimpse of the flow surface is fast disappearing.

As erosion continues to wear away the ridges on each side of this valley, this lava may eventually become a ridge-capping flow perched high above deep canyons, as have older flows on Forest and Lava Divides on the east side of Mount Baker. What was a valley bottom becomes a ridgetop through erosional inversion (Geologic Note 59 and accompanying figure).

About three-quarters of a mile below Schriebers Meadow (on USFS Road 13, 9.6 miles from the Baker River Road), look for roadcuts in red-orange basalt cinders. The small pieces of basalt are full of holes, which were once bubbles in the molten lava.

ANCIENT ROCKS ALONG THE PARK BUTTE TRAIL

Long history in a block

Most of Park Butte is an immense block of exotic gneiss in the Bell Pass Mélange (Chapter 3). You can best view these exotic rocks, called the Yellow Aster Complex, by taking a short cross-country hike off the Park Butte Trail. Leave the trail about two-thirds of a mile above Schriebers Meadow, where open fields of brush have just given way to open forest, and head southwest a few hundred feet through low scattered brush to reach Rocky Creek. The route is more or less across from a high, fresh rockfall scar on Survey Point. Cross Rocky Creek by way of logs, boulders, or wading as necessary. Be especially careful in high water.

Blocks from the rockfall are many and varied, but common is a light-colored, quartz-rich gneiss mixed with, or next to, layers of marble. These can be recognized by their gray color, rough, etched-looking surfaces, and the fact that they are easily scratched with a knife (Figure 64). The gneiss and marble in many blocks are

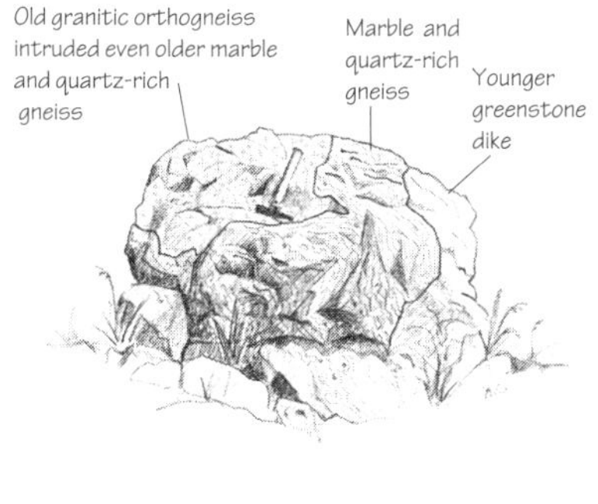

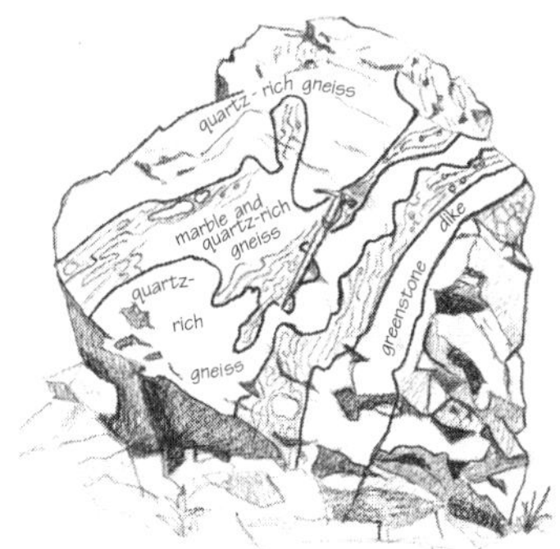

Figure 64. Blocks below Survey Point reveal the history of the Yellow Aster Complex. See also Figure 65.

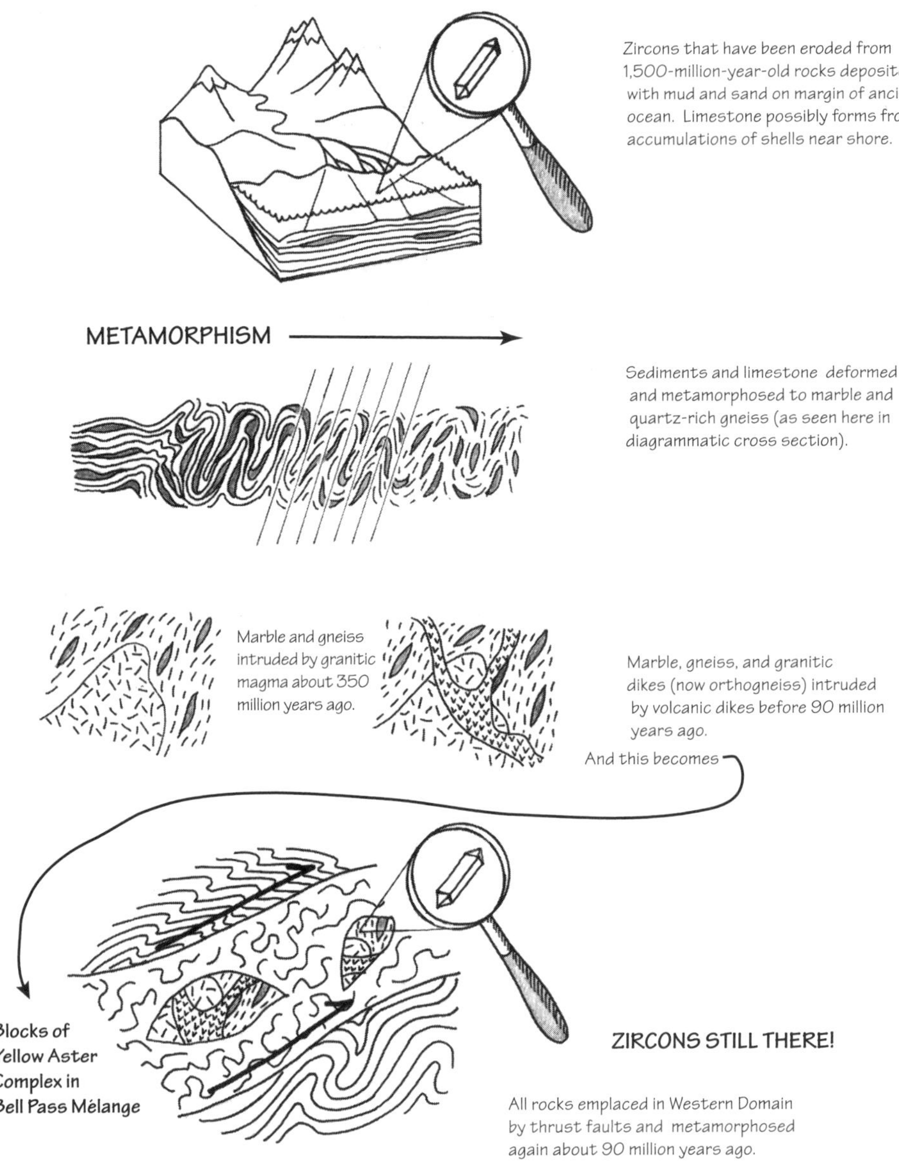

Zircons that have been eroded from 1,500-million-year-old rocks deposited with mud and sand on margin of ancient ocean. Limestone possibly forms from accumulations of shells near shore.
METAMORPHISM
Sediments and limestone deformed and metamorphosed to marble and quartz-rich gneiss (as seen here in diagrammatic cross section).
Marble and gneiss intruded by granitic magma about 350 million years ago.
Marble, gneiss, and granitic dikes (now orthogneiss) intruded by volcanic dikes before 90 million years ago.
And this becomes
Blocks of Yellow Aster Complex in Bell Pass Mélange
ZIRCONS STILL THERE!
All rocks emplaced in Western Domain by thrust faults and metamorphosed again about 90 million years ago.

Figure 65. Formation of the Yellow Aster Complex.

folded and distorted. Cross-cutting the gneiss and marble are a variety of igneous dike rocks, some white and coarse-grained, like granite, and some dark. The dark dikes are metamorphosed basalt, called *greenstone*. The gneiss layers commonly contain metamorphic minerals known to be mostly derived from shaley limestone.

These rocks and the marble indicate that the Yellow Aster gneisses here are metamorphosed from sedimentary materials that once accumulated in the ocean. Locally, the marble contains tiny gray, metallic-looking graphite flakes, the recrystallized carbon from plant or animal remains. Isotopic analyses of zircon crystals from related gneiss at Yellow Aster Meadows (Geologic Note 90) suggest that sediments were derived from an old continent with rocks that formed at least 1,500 million years ago (Precambrian). Zircons from the cross-cutting igneous dikes, on the other hand, indicate that the dikes crystallized about 350 million years ago (mid-Paleozoic), but after the marine limestones and shales were metamorphosed into marble and gneiss. The Precambrian zircons were probably sand grains, eroded from Precambrian rocks exposed in some ancient continent and then deposited sometime before 350 million years ago (Figure 65). Scott Babcock, professor at Western Washington

Figure 66. Panoramic view to the south from Railroad Grade moraine.

University and long a devotee of North Cascade geology, proposes, with a smile of creative imagination, that these rocks began life in the ancient supercontinent of Pangaea somewhere in the vicinity of today's Australia.

55

RAILROAD GRADE

View from atop a glacial moraine

Leave the Park Butte Trail after it reaches high meadows above a stiff switchback climb and take the Railroad Grade Trail to gain wonderful views of volcanic and glacial features (roughly 4.5 miles and 1,800 feet of climbing from the trailhead). Pass the designer camp sites and climb onto the moraine for the grand view; climb farther on up the moraine for even better views. The panorama from the moraine takes in the Easton Glacier, today well-retreated from its glory days as represented by the lateral moraine that provides this viewpoint. The relatively smoothly ascending ridge of the moraine, which inspires its name, mimics the surface of the glacier that deposited it (Chapter 7). Down the valley of Rocky Creek, the barren glacial debris gives way to brush and forest. Farther on down the valley (beyond the well-hidden parking area at Schriebers Meadow), a slight hump in the forested terrain is actually a small volcano—really a cinder cone—that gave vent to the Sulphur Creek lava flow that flooded the lower end of Rocky and Sulphur Creeks and continued on down to Baker Lake (Geologic Note 53).

Look up the mountain to the northwest to see the dark crags of Black Buttes, an ancient volcano that grew and was eroded before Mount Baker burst forth. Black Butte lavas are about 300,000 to 500,000 years old, whereas lavas from Mount Baker itself are probably all younger than about 30,000 years.

Look south to see mountains carved from rocks hundreds of millions of years older than Mount Baker lavas. Across Rocky Creek are the ridges of Loomis Mountain, carved from an old volcanic arc of the Chilliwack River terrane (Figure 66). These rocks range from about 150 to 400 million years old. Park Butte, in the near foreground on the west, is carved from a block of ancient gneiss and other rocks of the Yellow Aster Complex (Geologic Notes 54, 90). The block of ancient gneiss is incorporated in the Bell Pass Mélange and lies on a nearly horizontal fault above the Chilliwack River terrane. To the west in the distance are the peaks of the Twin Sisters massif, a huge piece of the Earth's mantle that is also caught up in the Bell Pass Mélange (Chapter 3). In good viewing light the orange hue of the dunite (Geologic Note 57) making up Twin Sisters is a striking contrast to other old rock in view, which is all black, gray, and brown.

56

SCOTT PAUL TRAIL

Volcanic rocks from Mount Baker

Take the Scott Paul loop trail for views of the Easton Glacier, its old moraines, a variety of volcanic rocks, and—in the distance—ancient rocks. East of the cable bridge over Rocky Creek (roughly 3.5 miles and 1,200 feet of climbing above Schriebers Meadow by way of the Park Butte Trail), look at the multiple hues in the glacier-borne rubble in the valley. Most of the red and gray rocks are relatively fresh andesite and dacite (Chapter 6; Figures 86 and 112) from Mount Baker. Rocks that are orange, brown, and white are altered lavas, probably derived from

Sherman Crater. Look for white cobbles of altered lava. On their broken surfaces, pale yellow coatings are native sulfur. Some rocks are pockmarked with bubble holes now filled with sulfur crystals.

Other geologic views from this trail are similar to those seen from the Railroad Grade moraine (Geologic Note 55).

57

THE TWIN SISTERS

Climbing on the mantle

Climbers bound for the summits of the Twin Sisters cannot but note the exotic feel and look of the orange-gold rock that they ascend. The whole Twin Sisters massif is made of ultramafic rock, mostly dunite, which came from the Earth's mantle. This huge slab of mantle rock was somehow incorporated into the Bell Pass Mélange (Chapter 3) when tectonic plates collided. The rock formed about 5 to 10 miles down in the Earth and was uplifted to its present position with remarkably little alteration of the deeply formed minerals. Dunite is mostly composed of olivine, but here also contains a small amount of pyroxene. The pyroxene crystals are a little more resistant to weathering than olivine, so they protrude from exposed rock surfaces as rough raisin-sized blisters. Some freshly exposed pyroxene crystals are emerald green. The rocks are thus rough and raspy, hard on hands and clothing, but pleasurably secure and firm under the boot. Black streaks and pods in the dunite are chromite (iron chromium oxide), a mineral mined elsewhere in the world for chromium.

ANDERSON CREEK ROAD (USFS ROAD 1107) AND BEYOND

58

SWITCHBACK ABOVE ANDERSON CREEK CROSSING

Exotic blocks in a rock mélange

The lower Anderson Creek Road cuts back and forth uphill through dirty gray and green rocks, mostly volcanic rocks of the Chilliwack River terrane. At Anderson Creek, the road crosses a significant fault and then zigzags up through the varied rocks of the Bell Pass Mélange (Chapter 3). On a switchback where the road loops north around a small glacial tarn about 9.2 miles beyond the Baker Lake dam, the road cuts through truly exotic rock. Exposed here is a dark streaky rock called *amphibolite,* which is mostly composed of hornblende and feldspar. The rock's thin laminations are commonly emphasized by tiny, elongated white blebs. Named the Vedder Complex after a mountain near Sumas, Washington, this amphibolite is metamorphosed basalt. What makes this particular outcrop interesting is that although it is seriously metamorphosed, the surrounding sedimentary rocks, argillite and chert, are not. From other Vedder localities, where considerable radiometric age dating has been done, geochronologists have found that metamorphic minerals in identical rocks crystallized about 250 million years ago (Permian). Chert, not too far away, contains Permian radiolarians that were dying and falling to the seafloor while the Vedder Complex was buried some 12 miles below the Earth's surface, recrystallizing to amphibolite. Later events mixed the amphibolite blocks into the chert and argillite, producing the mélange exposed in this roadcut.

59

ANDERSON CREEK ROAD OVERLOOK

A geologic potpourri

Where the road to Anderson Butte and Watson Lakes trailhead branches left (9.7 miles from the Upper Baker Dam), go straight about 0.5 mile to a bluffy roadcut of highly deformed greenstone and shale of the Bell Pass Mélange (Chapter 3; Plate 5C). The rocks are full of faults and fractures. From this spot—if the day is clear—look to the south and west and admire a variety of geologic wonders. (Many other spots on the road offer these same sights.)

In the middle foreground to the south, Welker Peak rises above the Welker Peak thrust fault. The peak itself is mostly chert in the Bell Pass Mélange. Below the fault are volcanic and sedimentary rocks of the Chilliwack River terrane (Chapter 3; see Plate 2).

To the northwest, Mount Baker dominates the scene. The mountain and its old-rock foundation rise immense from the floor of the broad Baker River valley. On the near side of the volcano, glaciers point their tongues down valleys eroded between sharply sculpted old lava flows, called *cleavers.* These lava-flow caps on today's ridges were once hot lava floods in valley bottoms, since left high by ero-

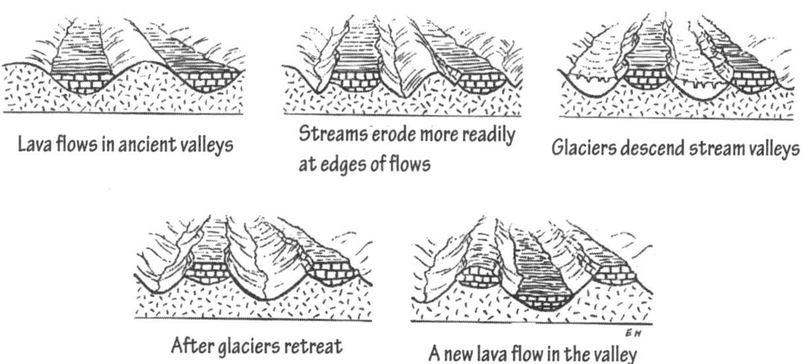

Figure 67. Erosion along the sides of a valley-filling lava flow leads to an inversion of topography.

sion. Such topographic reversal is common on the flanks of volcanoes, where a valley bottom lava flow displaces creeks to the sides of the flow. In time, the newly routed creeks erode valleys that flank the original valley-bottom flow (Figure 67; Geologic Notes 53, 77). Recent studies of some ridge-cap flows indicate they may have been confined by glacial ice. The lava may have flowed out between valley glaciers, and if so the topographic inversion might not be so marked because much of the valleys were already there.

Looking southwest down the Baker River, conjure up a vision of a huge glacier filling the valley below. To view this scene from this vantage in the Late Pleistocene (about 17,000 years ago) an Ice Age visitor would have had to stand on the ice surface some 800 feet above the road. A vast sea of ice would stretch out to the south. High peaks of the main Cascade Range would rise above the ice to the east, and on a very clear day, the dark peaks of the Olympic Mountains would show on the southwest horizon. The ice surface would have extended from 5,000 feet here to about 3,000 feet on the northeast flank of the Olympics. More likely, everything would have been obscured by blowing snow.

At one stage during glacial retreat, the outwash from Cordilleran ice in the lower Skagit River valley built a thick fan of gravel and sand at the margin of the melting glacier. The small hills in mid-distance toward the southwest (Figure 68) are remnants of the outwash fan, which sloped down to the northeast, away from the ice.

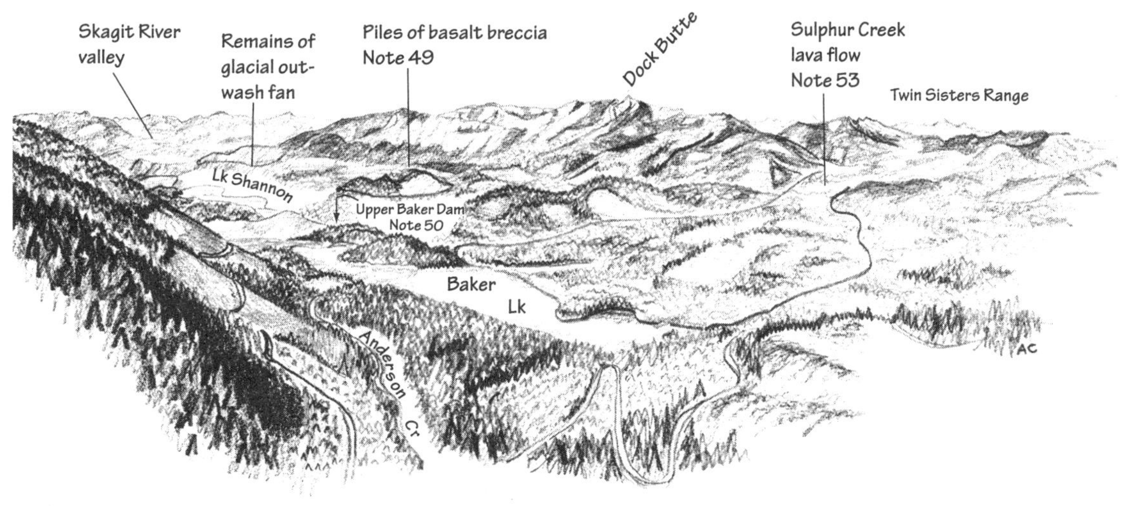

Figure 68. View to the southwest down the Baker River valley from Anderson Creek Road.

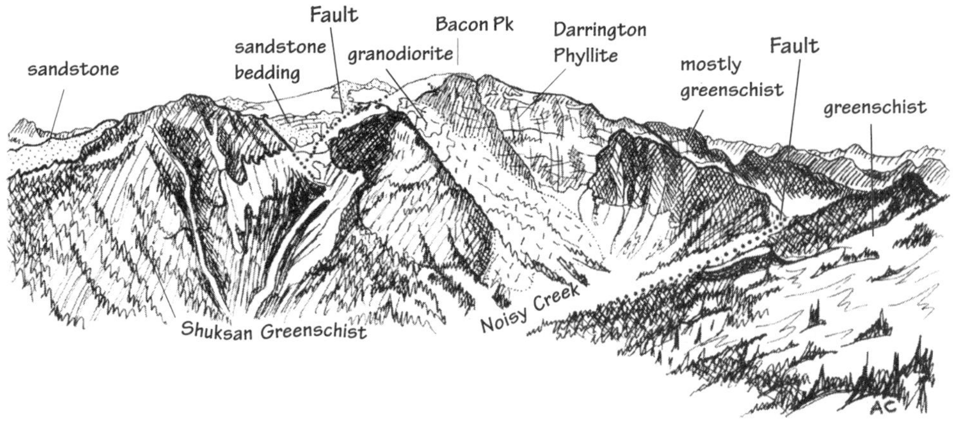

Figure 69. View from the true summit of Anderson Butte looking east to Bacon Peak and sandstone beds lying on rocks of the Easton terrane. A small arc-root pluton of granodiorite intrudes the older rocks.

60

ANDERSON BUTTE

A view of Bacon Peak and its cap of Eocene sediments

From the summit of Anderson Butte (an easy hike of about 1.8 miles and 1,000 feet of climbing from the Anderson Lakes trailhead), the traveler has great views of mountains carved from rocks of the Western Domain. Almost directly east is Bacon Peak (Figure 69), a massif carved from rocks of the Easton terrane (Shuksan Greenschist and Darrington Phyllite) and capped by 50-million-year-old sandstone and conglomerate of the Eocene extensional event. During the Eocene extensional event (Chapter 5), faulting and stretching of the crust produced low areas—valleys—that were filled up with sand and gravel by streams (Plate 8B). Geologists do not know how much of the immediate area was once covered by the stream deposits. Most have been eroded away except for patches like that on Bacon Peak, which were faulted down into the older, harder rocks and therefore somewhat protected from erosion.

WEST SIDE APPROACHES

CASCADE RIVER DRAINAGE

CASCADE RIVER ROAD AND BEYOND

61

ELDORADO PEAK VIEWPOINT

Metamorphosed conglomerate and the Chelan Mountains terrane

At 6.9 miles from Marblemount, along the Cascade River Road, look for a turnout on the right (south) and a well-constructed old-style guardrail of small logs and aggregate posts. From the turnout, look upvalley to see Eldorado Peak (8,868 feet) at the head of the Marble Creek cirque to the left and Hidden Lake Peaks on the right. Both of these mountains are carved from orthogneisses. In the foreground the view also includes a bend in the Cascade River valley, where it meets the Entiat Fault (Geologic Note 65). The river has adjusted its course along the more easily eroded rocks of the fault zone.

The roadside outcrops here expose the Cascade River Schist of the Chelan Mountains terrane. Common in the Cascade River Schist is metamorphosed conglomerate, which can more or less be examined here, but only the most determined petrophile will see much. Lichens and road dust obscure large boulders of light-colored plutonic igneous rock in the dark mica schist—once the sandy matrix of the conglomerate (Geologic Note 71). Walk a few hundred feet back down the road, looking high on the outcrop, above the worst dust coating, to see vague, rounded light-colored patches which are the boulders. Rounded knobs protruding from the outcrop are also boulders.

SOUTH FORK CASCADE RIVER

A deep U-shaped canyon

Hikers who venture a few miles up the South Fork of the Cascade River Trail experience a truly wild and impressive glaciated canyon. The U-shape of the South Fork canyon is readily apparent at any opening in the dense woods where the steep canyon walls are visible. Steep cones of talus have modified the classic U-shape (Figure 36). While impressive, the South Cascade Glacier, viewed only after hiking many miles of crude trail up the canyon, is but a feeble reminder of the mighty ice river that carved these precipitous walls.

GILBERT CABIN

Mines and avalanches

The site of Gilbert Cabin, a prospector's base for a mine high on the north side of Johannesburg Mountain across the valley, is located where the Cascade River Road makes a couple of sharp hairpin turns in its climb out of the North Fork valley to reach the parking area below Cascade Pass. At one time a cableway was rigged to the mine from this site. Old reports indicate that many tons of lead, silver, and gold ore were removed, at first by trail and then by aerial tram.

From Gilbert Cabin the road climbs steeply up across a cirque headwall and crosses several roaring cascades: Boston, Morning Star, Midas, and Soldier Boy Creeks, all named for mineral prospects along them. These streams pour down alder-filled brush chutes that mark snow avalanche paths. Typical of the alpine scene in the North Cascades, these avalanche chutes play an important erosional role. Huge masses of snow sliding down them prevent trees from growing and carry rock and soil to the main valley below. The avalanche debris builds fans where the steep slope meets the gentler valley bottom. In some valleys, broken trees and the brush pattern show where one or more avalanches rushed down one side of the valley and up the opposite side as well. In such areas almost all man-made structures—cabins, mine buildings, bridges—are eventually overrun by avalanches unless very carefully placed to avoid their paths. Visitors may encounter remnants of avalanche snow on or along the road at these creeks.

64

CASCADE PASS TRAILHEAD

Icefalls and ancient volcanic blasts

In the North Cascades, few places can be reached by car that offer a close view of alpine rock and glacier. The parking area at the Cascade Pass trailhead is one such place. Across the valley from the parking area, three spectacular peaks—from left to right (east to west), The Triplets, Cascade Peak, and Johannesburg Mountain—rise almost vertically some 3,500 to 4,600 feet. Blocks of ice from the hanging glaciers on Johannesburg often fall with a roar down the dark cliffs.

Johannesburg Mountain (8,200 feet) and much of Cascade Peak (7,428 feet) are carved from schist and metaconglomerate of the Cascade River Schist, a part of the Chelan Mountains terrane (Chapter 3). The Triplets (7,240 feet) are made of schist breccia—a jumble of shattered schist fragments cemented by quartz. Red-stained rock east of The Triplets and at their base reveals the cause of the shattering, an igneous tonalite, part of the very large Cascade Pass dike, which rose along a northeast-oriented crack in the schists (Figure 70). The Cascade Pass dike is over 9 miles long and almost a mile wide in some places. This was some crack. The dike, an arc-root pluton of the Cascade Pass family, is 18 million years old (Miocene) (see Plate 2; Figure 30). It narrows considerably where it goes over ridges, indicating that it has a bladelike shape in cross section.

The shattered rocks of the Triplets reveal a place where gases and liquids escaped the crystallizing magma with explosive force, probably through a volcanic vent to the surface. Most of the mineral deposits in the area are associated with the Cascade Pass dike and probably formed when hot watery solutions carrying metals from the magma of the dike permeated the surrounding rocks (Geologic Notes 26, 127). For the best view of the dike, pause on the final stretch of trail to Cascade Pass (Geologic Note 69).

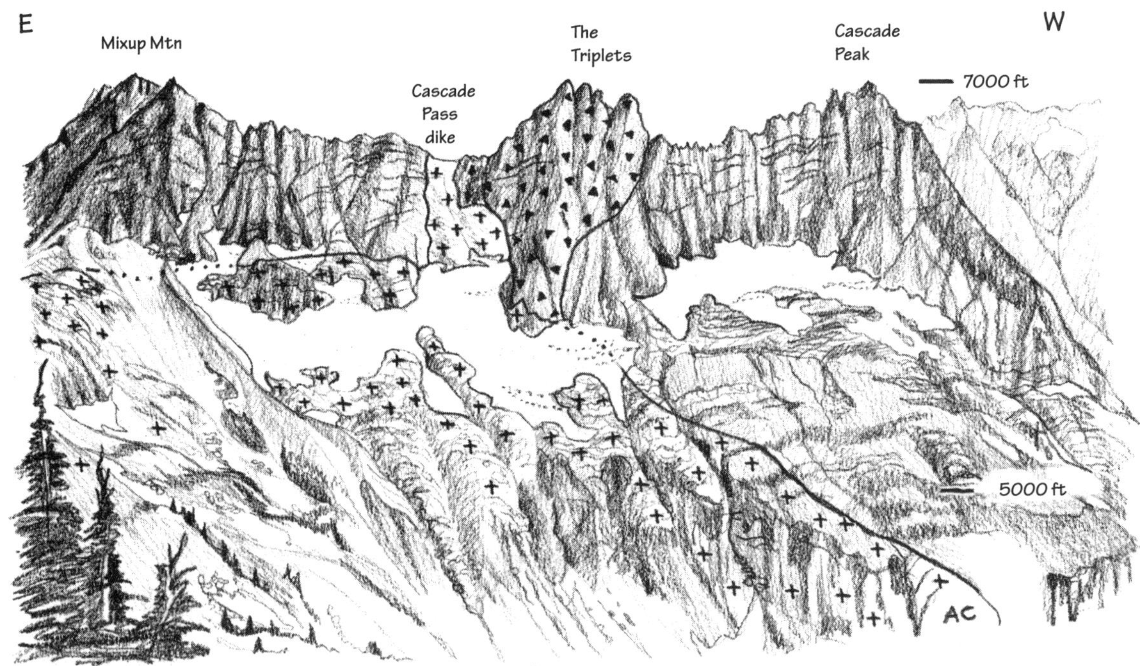

Figure 70. View from the Cascade Pass Trail of the Cascade Pass dike where the dike crosses the ridge between Mixup and Johannesburg Mountains. Crosses show the location of the tonalite of the dike. Triangles show the explosion breccia on The Triplets. Dots indicate contacts hidden under ice and moraine.

TRAILS REACHED FROM THE CASCADE RIVER AND BEYOND

65

LOOKOUT MOUNTAIN TRAIL

Schist and views of the Entiat Fault

Fortunate hikers who attain the summit of Lookout Mountain (5,693 feet; about 6 miles and 4,500 feet of climbing from the signed trailhead on the Cascade River Road) on a beautiful day can see from this one viewpoint many of the geologic features described in this book. Rocks exposed along both the trail and summit ridge of the peak are fine-grained biotite schist of the Cascade River Schist. These rocks were formed by metamorphism of shale, volcanic ash, and sandstone deposited in a submarine fan at the toe of a volcanic arc. Erosion-resistant metamorphosed conglomerate holds up the steep south face of Lookout Mountain (Geologic Notes 61, 71), but luckily for the trail builders, who habitually avoid hard rock, this metaconglomerate is not present along this ridge. The cobbles in the original conglomerate were eroded from uplifted arc-root plutons (Chapter 3). The Cascade River Schist, born of the volcanic arc, is here separated from rocks of deeper ocean birth, the Napeequa Schist, by the Entiat Fault, which runs through the notch between Lookout Mountain and the ridges around Monogram Lake (Geologic Note 66; Chapter 5). The broad valley of the Cascade River stretching to the south has been eroded out along the fault (Figure 71). The trace of the fault is difficult to see to the north, but continues into Bacon Creek, where it intersects the Straight Creek Fault (Geologic Note 2; see Plate 2), which runs north-south but is equally obscure from this vantage.

The Entiat Fault stretches from its northern end in Bacon Creek southeast to Wenatchee, a distance of roughly 100 miles. Geologists are still arguing about the overall history of this obviously major fault, but here careful study of the rocks on either side has shown that the Cascade River Schist west of the fault was metamorphosed at much shallower depths than the Napeequa Schist east of the

Figure 71. Looking south from Lookout Mountain. The Cascade River takes a southeast-trending jog along the more easily eroded rocks of the Entiat Fault Zone. The drawing, made from photographs, makes the fault look curved, although it is really straight. Most of the rocks in view are of the Chelan Mountains terrane and stitching plutons.

fault. The block on the east side has risen about 14 miles relative to the block on the west side!

66

MONOGRAM LAKE

Uplifted schists and granitic rocks

After climbing up some 4,300 feet to reach Monogram Lake (about 5.5 miles from the Cascade River Road), mountain travelers might be comforted to know that the surrounding rocks have gone up even more than they have. About where the Monogram Lake Trail joins the Lookout Mountain Trail (after what seems like a million switchbacks), hikers leave schists metamorphosed at relatively shallow depths, cross the Entiat Fault, and resume climbing through igneous rocks of the Marble Creek pluton, which intruded the schists and was metamorphosed at great depth. At and near Monogram Lake, hikers will find outcrops of granitic rock interspersed with the Napeequa Schist of the Chelan Mountains terrane, as well as ultramafic rocks. The Marble Creek pluton in this area was more like a network of dikes and sills (Geologic Note 7) than a continuous mass of magma. It is now distorted into streaky layers. The chemical composition of the minerals in these rocks, as well as the crystallization of epidote from the magma (look for minute straw-yellow "worms" infesting dark biotite splotches in the rock), shows that they crystallized some 16 to 19 miles deep in the Earth. Studies in the lab, where petrologists cool artificial magma in pressurized containers (mimicking depths in the crust) to see what minerals form in the artificial rock, give geologists a pretty good idea of what minerals crystallize from magmas at various pressures at depth in the crust.

67

HIDDEN LAKE PEAKS STOCK

A piece of the mantle high on a ridge

Climb up to the lookout at the summit of Hidden Lake Peaks for views and a good look at the Hidden Lake Peaks stock (about 4 miles and 3,300 feet of climbing). A stock is a relatively small granitic pluton. This one intruded the Napeequa Schist of the Chelan Mountains terrane some 75 million years ago. The stock has been recrystallized by the heat of metamorphism but escaped being squeezed out like its neighbor, the Marble Creek pluton (Geologic Note 66). The dark mineral in this rock is the mica, biotite.

For a look at really interesting rocks of the Chelan

Mountains terrane, descend with ease along the ridge northward toward the saddle at the head of Sibley Creek. At the first notch separating a gendarme of schist from the granitic rock is a sharp contact between the white granitic rock of the stock and adjacent brown schists. Here the magma of the stock stopped against the schist. Recall from Chapter 3 that the Napeequa Schist formed from oceanic shales, basalts, and ribbon cherts. In the lowest saddle find a dark, messy-looking green rock with shiny broken surfaces. This is a pod of ultramafic rock, one of many in the Napeequa Schist. The original pods were slivers of the Earth's mantle, probably incorporated into the original oceanic sediments and basalts of the Napeequa Schist along the continental margin during subduction (Chapter 2). The original high-temperature, high-pressure minerals that formed in the mantle (olivine and pyroxene) have been mostly metamorphosed to talc and serpentine. From the former mineral, talcum powder is made, and the surfaces of the rock may feel soft, dry, and soapy (as in soapstone).

68

A MARBLEMOUNT PLUTON

Ancient arc plumbing

Hikers following the tortuous South Fork of the Cascade River Trail (about 3 miles from the trailhead), or climbers on Le Conte Mountain, will find a brownish green, granitic-looking rock. In places it looks like a dirty gneiss. This rock is a Marblemount pluton, one of a family of metamorphosed igneous plutons aligned in a belt stretching from north of Marblemount to the southeast, beyond Holden (see Plate 2). This messy-looking, metamorphosed tonalite orthogneiss is a root pluton of an ancient volcanic arc that erupted about 220 million years ago (Triassic; Chapter 3). In spite of its crumbly look, the orthogneiss of the Marblemount plutons holds up a number of prominent peaks in addition to Le Conte. To the southeast it also underlies Sentinel Peak, Old Guard Peak, Bonanza Peak (Geologic Note 123), and Dumbell Mountain.

CASCADE PASS TRAIL AND BEYOND

69

CASCADE PASS

Recording the ages of rocks using the radiometric argon clock

The high alpine views, not to speak of the rocks, are well worth the short climb (3 miles, 1,750-foot elevation gain) on gentle trail to Cascade Pass (5,360 feet). Hikers on the trail can see white outcrops of granitic rock on either side of the pass. This rock is tonalite of the relatively young Cascade Pass dike of the Cascade Pass family (Geologic Note 64 and accompanying Figure 70). The dark, conspicuous minerals are hornblende and biotite. The lighter-colored minerals are feldspar and quartz.

The age of the Cascade Pass dike has been determined by potassium-argon isotope analysis of its minerals. One isotope of the element potassium decays radioactively to argon gas. As the rate of decay, that is, rate of change from the radioactive isotope to argon, is constant, measuring the amounts of the potassium and argon in the mineral will reveal its age. The minerals hornblende and biotite both contain potassium. They can be readily separated from the rock, and their potassium and argon content can be measured with great accuracy. The potassium-argon method of dating rocks is very useful for many geologic problems because many rocks contain potassium-bearing minerals. Unfortunately, argon is a gas and a noble gas at that; it does not combine with other elements in the mineral and only remains if it is trapped in the crystal lattice. If the rock gets reheated, much of the argon may escape. Because the argon escapes from the crystals of hornblende at a slower rate than it does from biotite, the geochronologist knows that if the radiometric ages of the two minerals are the same, then chances are that the rock has not been significantly reheated. In the example discussed here, the hornblende and biotite have the same ages, indicating that argon has not escaped and has remained in both minerals for 18 million years.

Just below the west side of the pass, hikers have an excellent view northwest to snowy Eldorado Peak (8,868 feet), a massive, roof-shaped mountain made of the Eldorado Orthogneiss. Hikers who want to view the orthogneiss up close can climb north, up onto Sahale Arm, or hike on towards Stehekin (Geologic Note 128). The pluton which became the Eldorado Orthogneiss intruded the rocks of the Chelan Mountains terrane about 90 million years ago.

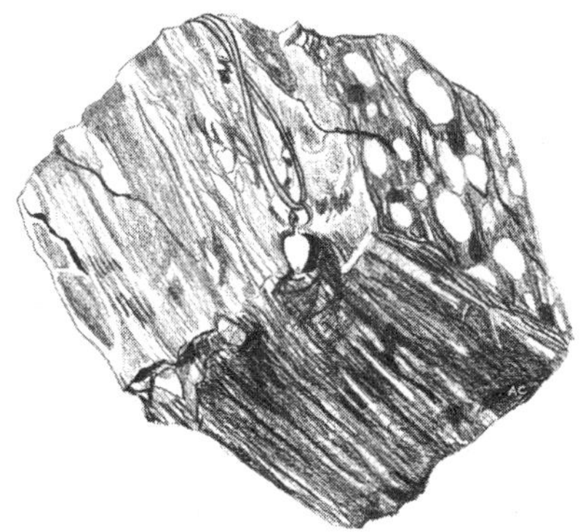

Figure 72. Block of metaconglomerate showing pebble shapes on one side and, on the other sides, mostly streaks that are the flattened pebbles viewed on edge.

Because this pluton has had a long history of reheating, geologists could not use potassium-argon dating to determine its age. Instead, they analyzed uranium and its decay product, lead, both found in zircon contained in the orthogneiss (Geologic Note 4).

Looking east from Cascade Pass to distant ridges beyond the Stehekin River valley, hikers can see mostly orthogneisses of the Skagit Gneiss Complex (Chapter 3).

(For information on rocks south of Cascade Pass, see Geologic Note 129 and Figure 101 and for geologic points of interest east of Cascade Pass, see Geologic Notes 127 and 128.)

70

SAHALE ARM

Rocks of the grand view

Hikers who climb from Cascade Pass to the upper reaches of Sahale Arm, southwest of Sahale Peak, get a grand view of Magic Mountain and the summit ridge of Johannesburg Mountain (Figures 70, 101). Across Pelton Basin, to the south, the squat brownish Pelton Peak on the east (left) is underlain by the Cascade River Schist of the Chelan Mountains terrane. The same rocks underlie upper Sahale Arm (see Plate 2). The sharply bladed peak west of a broad notch, filled with the Yawning Glacier, is Magic Mountain, made of the Magic Mountain Gneiss (Geologic Note 72). The notch and the broad shelf bearing the Cache Col Glacier to the west are eroded along faults. On the far west (right), Johannesburg Mountain's dark fluted cliffs, with their clinging glaciers, are held up by metaconglomerate and schist, cooked hard and made resistant to erosion by the Cascade Pass dike, which underlies the lower slopes of Cascade Peak and The Triplets (Geologic Note 64).

71

TRAIL TO CACHE COL

Metaconglomerate and other schists

A mountain climber's path leads south and up from Cascade Pass. Scramble up it to find many white outcrops of tonalite of the Cascade Pass dike (Geologic Notes 64, 69). Beyond the dike, the trail crosses scree and talus of brown rock of metamorphosed conglomerate of the Cascade River Schist of the Chelan Mountains terrane (Chapter 2; see Plate 2). In most of the blocks, pebbles look like roundish or elongated white patches in a streaky, somewhat swirled greenish matrix. In many rocks the pebbles have been stretched into long rods, and these parallel rods look like white layers in the rock (Figure 72).

Experienced hikers with ice axes can continue up steep scree (often covered with snow) to a rocky shoulder at the side of the Cache Col Glacier, where wonderful glacier-polished outcrops of schist also reveal the sedimentary origin of these rocks. Look for alternating layers of black schist and light-colored schist. The dark layers were beds of carbonaceous shale and sandstone; the light-colored layers were less shaley sandstone. The black schist contains many tiny red garnets (Geologic Note 108). The towering walls of Mixup Peak (7,440 feet) above are of Magic Mountain Gneiss (Geologic Note 72). Do not go onto the glacier unless you are prepared with proper equipment and experience for glacier travel. A hidden crevasse or the *bergschrund* could swallow you up (Figure 37; Geologic Note 83).

72

CACHE COL

A Magic Mountain Gneiss

When climbers bound for the high country south of Cascade Pass reach the pass at Cache Col, they are rewarded by a great view southward of Mount Formidable and its jumbled icefalls. The rock at the pass is iron-stained and crumbly, as well it should be in a fault zone (Plate 4C). Above the Middle Fork of the Cascade River, a swale in the talus trends southward across the basin, a depression eroded out along the crushed rocks of the fault zone. A little way from the col on either side, outcrops are fresher and composed of well-layered Magic Mountain Gneiss, which is made up of light-colored gneiss and dark green schist. This rock is unusual in containing minerals that are not normally associated with gneiss. Gneiss is usually considered an advanced metamorphic rock, meaning that it is generally well recrystallized at the high temperatures and pressures found deep in the Earth's crust (Chapter 2). The minerals contained in these layers of Magic Mountain Gneiss (one of which is chlorite, a green and somewhat greasy-looking mica) formed at low pressures and temperatures, resulting in an unusual low-temperature orthogneiss. This metamorphic oddity formed when a Marblemount pluton was squeezed and recrystallized at a relatively shallow depth in the Earth's crust.

Across the valley of the Middle Fork of the Cascade River, the upper part of Mount Formidable is made of Magic Mountain Gneiss, but the lower parts of the mountain are of schist derived from sedimentary rocks.

WEST SIDE APPROACHES:

NORTH FORK OF THE NOOKSACK RIVER DRAINAGE

MOUNT BAKER HIGHWAY (STATE ROUTE 542)

73

CHURCH MOUNTAIN LANDSLIDE

A catastrophic event

The traveler driving up the Mount Baker Highway for the first couple of miles beyond the town of Glacier sees little evidence of the huge landslide that rushed onto the valley floor here. Erosion, stream deposition, and the forest help conceal the hummocky surface of the slide, which fell from Church Mountain to the north. Based on radiocarbon ages of buried wood (see Glossary), the landslide crashed into the valley about 2,400 years ago. Where the highway ascends a long hill east of Glacier Creek, it is climbing out of the young, eroded valley of Glacier Creek onto the thick debris of the landslide. A condominium development in the valley is tastefully placed around the humps and bumps of the landslide's surface. River rafters on the North Fork of the Nooksack here dodge huge boulders eroded out of the landslide. Geologists look for a special event as the trigger for such a large slide—at least hoping for a special circumstance makes them feel more secure in these steep-walled valleys—and suggest that an earthquake might have been the cause.

74

NOOKSACK FALLS

The resistance of ancient volcanic rocks

A short drive off the Mount Baker Highway (7.9 miles from Glacier) leads to a view of Nooksack Falls, where the Nooksack River plunges over a cliff of ancient volcanic rock. Ancient indeed; the rock is some 180 million years old and represents arc volcanoes that erupted in a Jurassic ocean flooded with mud and sand. These are rocks of the Nooksack terrane (Chapter 3). Most of the sedimentary beds of the Nooksack lie above the falls here, layer after layer making up the cliffs of Excelsior Ridge to the north and Skyline Divide and other ridges to the south (Chapter 2; Geologic Note 84).

The present stair-step in the valley floor separates

an upper wide valley from a lower wide valley. Since the water rushing over the falls erodes the step back, the falls will eventually work their way upriver. Progress is slow in the hard volcanic rocks but will be quicker when the step reaches the softer shales and sandstones upstream (Chapter 7). Interestingly enough, the smaller tributary of Wells Creek, which joins the Nooksack in a gorge just below the falls, appears to have already sawed through the hard lump of volcanic rock. Actually, it is eroding a thick, but fortuitously located softer layer of sedimentary rock in the volcanic pile and therefore has managed to cut its way down faster than the main Nooksack River.

75

ABOVE NOOKSACK FALLS

A valley-puddling lava flow and its joints

About one-half mile east of the turnoff to Nooksack Falls, a large black cliff comes into view on the south side of the river. The conspicuous columnar jointing is characteristic of lava flows and forms when lava cools and shrinks. The shrinkage around evenly spaced centers is similar to the familiar shrinkage of mud, which also produces cracks in a polygonal pattern (Figure 73). This thick flow of andesite is not the ancient volcanic rock of Nooksack Falls, but a geologic youngster that erupted only 200,000 years ago. From some unknown vent, thick lava flows filled the canyon of the ancestral Nooksack North Fork. Eventually, these remnants might become a ridge cap as erosion inverts the topography (Geologic Notes 59, 77). Just below the Mount Baker Ski Area, look for more columnar joints in other lava flows cut by the highway.

The Nooksack River has left little valley floor between the steep canyon side of ancient volcanic rocks and this young pile. The highway and river make a tortuous curving route between them. The old rocks of the Nooksack volcanic arc along the north side of the highway here are stained brown and orange by iron from the mineral pyrite. The hot solutions that brought the pyrite into the rocks also brought in quartz and feldspar veins containing gold and silver. Miners removed much gold and silver in the early 1900s from the Excelsior Mine across the river.

Figure 73. Columnar joints in lava flow. Cooling lava shrank in the direction indicated by the arrows.

76

MOUNT BAKER SKI AREA

Summarizing the Western Domain

On a nice day the views from the Mount Baker Ski Area just about summarize the geology of the Western Domain. Across the small glacial tarns to the west, the ramparts of Mount Herman are made of volcanic rocks of the Chilliwack River terrane. Sedimentary beds between these ancient lava flows contain radiolarians (Geologic Notes 30, 103) that indicate they are about 255 million years old (Permian). To the north and to the west across the North Fork of the Nooksack River is Yellow Aster Meadows, which is underlain by the Yellow Aster Complex, a slice of old continent in the Bell Pass Mélange (Chapter 3; Geologic Notes 54, 79, 87, 90). To the northeast is Goat Mountain, which is composed of Darrington Phyllite of the Easton terrane. To the east is Mount Shuksan, a breathtaking mountain citadel that has appeared on innumerable calendars. Although familiar to many people for its alpine beauty, Mount Shuksan has even greater notoriety among geologists for being the type area of the Shuksan Greenschist, a rock derived from metamorphism of ocean-floor basalt (Geologic Note 80). All these older rocks have been draped by the geologically young lavas found underfoot here and on Table Mountain to the south.

77

ARTIST POINT

Old volcanoes

At Artist Point the geologic views are even better than from the ski area below. The dark prow of Table Mountain dominates the scene to the west. The parking area is on the same dark andesite that forms Table Mountain, but to really appreciate this old lava ridge cap, take the trail around it by way of the Galena Chain Lakes or climb over it. The top is tablelike because it reflects the original surface of

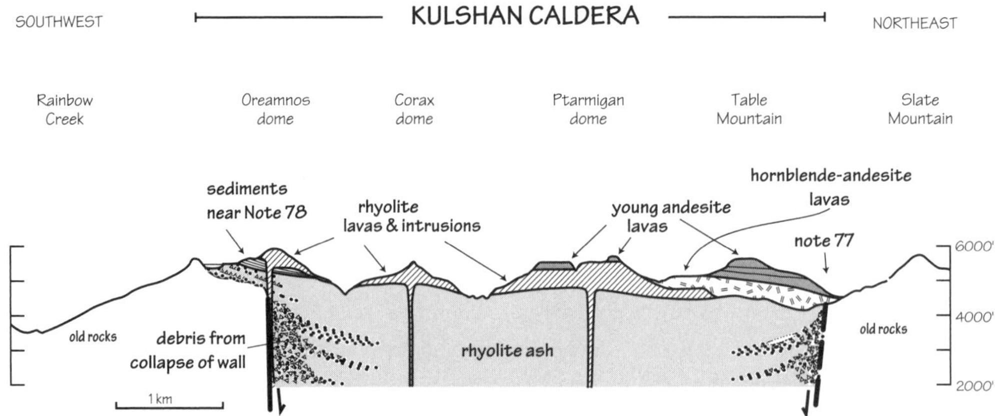

Figure 74. Kulshan Caldera filled with rhyolite ash (a cross-sectional view from the southeast; after Hildreth, 1996).

the thick lava flow. The Table Mountain flow in particular is so thick that geologists think it must have flowed into a constraining canyon in which the molten rock formed a lava lake. Because of the wild, crazy columnar joints that occur along the sides of Table Mountain, volcanologist Wes Hildreth suggests the flow may have been constrained by glacial ice! Other flows of about the same age underlie the ski area and make up parts of Ptarmigan Ridge stretching off to the southwest.

After these valley-filling lava flows solidified, streams and/or glaciers worked on the edges of the flows more efficiently than in the center and eventually left the remnants as ridgetops (Geologic Notes 53, 59). Erosion produced this marvelous topographic inversion after the lavas erupted about 300,000 years ago. These are not Mount Baker lavas; they erupted from an older volcanic vent located somewhere near the northeast flank of the present Mount Baker cone.

Under the dark lavas of Table Mountain, and forming white cliffs above Swift Creek, are the older volcanic deposits of Kulshan Caldera (Geologic Note 78). The edge of this large volcanic depression is more or less directly beneath the parking area. The caldera, about 2.5 miles across, formed and was filled with volcanic *tuff* (the rock formed from volcanic ash) about 1.1 million years ago, when the magma chamber beneath it erupted, and its roof collapsed. Similar volcanic calderas, such as the one filled by Crater Lake in Oregon, have produced huge volumes of ash in cataclysmic eruptions (Geologic Note 3). Volcanologists have identified ash deposits from the Kulshan volcanic eruption as far away as southern Puget Sound. The caldera itself is filled with over 3,000 feet of volcanic ash (rhyolite tuff) from such an eruption (Figure 74).

ROADS AND TRAILS REACHED FROM THE NORTH FORK OF THE NOOKSACK RIVER AND BEYOND

78

PTARMIGAN RIDGE

A lake in ancient Kulshan Caldera and younger lavas of Coleman Pinnacle

Hikers on the scenic Ptarmigan Ridge Trail travel across the eroded top of Kulshan Caldera (Geologic Note 77) and younger lava flows that erupted onto the deposits that fill the caldera. The caldera filling is mostly white to yellowish-white rhyolite tuff (volcanic ash), which is difficult to recognize with the naked eye. Peering through the microscope, however, the enhanced eye can see fragments of altered volcanic glass.

Where the trail skirts the south side of Coleman Pinnacle and the view opens up into the valley of the Rainbow Glacier coming down Mount Baker (about 3.5 miles from Artist Point), the trail crosses mostly volcanic ash that fell or washed into a lake. The caldera depression, after filling mostly with erupted ash, at least partially filled with water, creating a scene very much like Crater Lake

Figure 75. Highly deformed, metamorphosed banded chert (of the Bell Pass Mélange) in a block along the trail to Lake Ann.

in Oregon, but a million years older. The thin water-laid beds of these deposits can be examined in greater detail by hiking south along the ridge south of Coleman Pinnacle. A hike farther southeast to a broad rounded summit (southeast of Coleman Pinnacle; elevation 5,847 feet on topographic maps) reveals the exhumed root of a rhyolite dome—an extrusion of thick, pasty lava—not too different from the dome that formed in the crater of Mount St. Helens after it erupted in 1980.

Coleman Pinnacle itself consists of younger lava that erupted on top of the caldera deposits. This andesite lava is about 300,000 years old and is characterized by its gray color dashed with black hornblende crystals. The trail to this point crosses many outcrops of this gray lava.

79

LAKE ANN TRAIL

A geologic panorama of the last 225 million years

Perhaps one of the most rewarding scenic and geologic hikes in the Mount Baker area is the Lake Ann Trail. At the divide between Swift Creek and the cirque of Lake Ann (about 3.8 miles from the trailhead), pause to take in the geologic scene. A slow pirouette will bring into view geologic events reaching back 225 million years. An even better view is afforded by climbing a short ways on a primitive trail up the ridge to the south.

From either viewpoint, start with the view west, back down the trail. Shuksan Arm, the ridge north of the trail on the right, is held up by the oldest rocks in view, sedimentary rocks of the Chilliwack River terrane (Chapter 3). Closer to you on the same ridge, but up the slope to the northwest, a slab of chert was thrust over the sedimentary rocks along a fault. This chert is part of the Bell Pass Mélange. Blocks of the mangled recrystallized chert are scattered along the trail to Lake Ann in the basin to the west of this point. Look for white blocks made up of broken lenses and layers of quartz. Some blocks look like weathered, coarsely grained wood (Figure 75). These are metamorphosed, finely bedded chert and shale (ribbon cherts) of the ocean bottom. On Shuksan Arm, the mangled chert and shale are bound on the east by the Shuksan Thrust Fault which cuts across the ridge almost directly north of this view point. Above and east of the thrust is a slab of phyllite, overlain, going eastward, by the Shuksan Greenschist, which makes up the main bulk of Mount Shuksan. The phyllite and greenschist are of the Easton terrane, and geologists think these rocks began in the ocean as shale and basalt about 150 million years ago (Jurassic). The same sequence of tilted layers also occurs on ridges to the south. You are standing here in a bite out of the edge of a stacked sandwich of thrust sheets which are tilted down to the southeast (Figure 76).

A lot of geologic time is not represented by the views here because the next youngest rock that can be seen is the granitic rock of the Lake Ann stock, a small pluton that intruded the older rocks about 2.2 million years ago (Pliocene). The granitic rock, part of the Cascade Volcanic Arc, is underfoot as well as exposed on nearby slopes. A lot of things happened between 150 and 2.2 million years ago, including descent of the Easton terrane to great depths in the Earth, uplift of the terrane, travel to somewhere in western North America, assemblage with other terranes by thrust faulting, and eventual arrival of the stacked terranes at a place on Earth that we now call western Washington.

To the west, across the headwaters of Swift Creek, are white cliffs carved from ash beds of Kulshan Caldera. The tuffs of these cliffs are 1.1 million years old, even younger than the Lake Ann stock. The black cliffs of

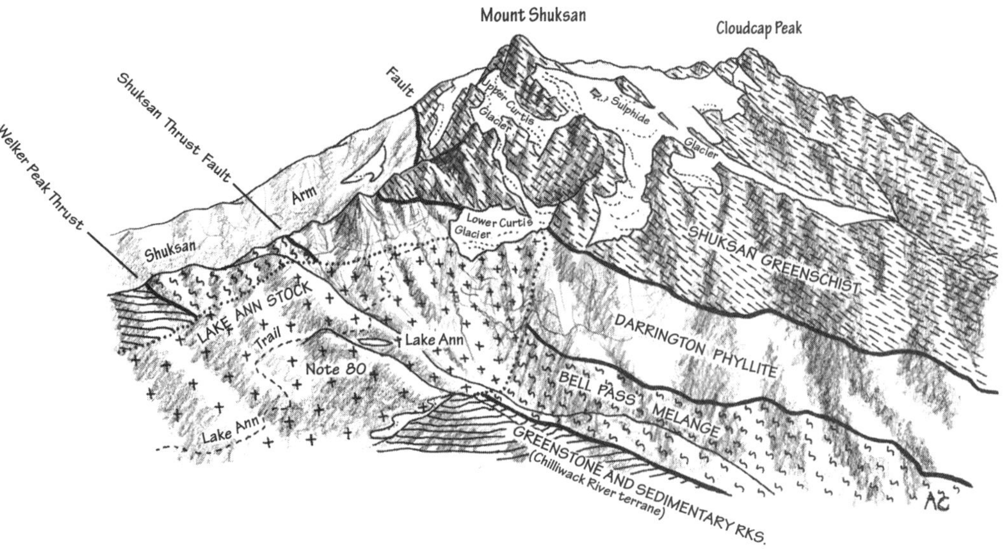

Figure 76. View of Mount Shuksan from the west showing stacked sandwich of thrust plates tilted down to the southeast. About 2 million years ago the magma of the Lake Ann stock welled up into the stacked thrust plates. The Shuksan Greenschist and Darrington Phyllite make up the Easton terrane.

Table Mountain represent the next youngest event, eruption of thick lavas about 300,000 years ago (Figure 77; Geologic Note 77).

80

MOUNT SHUKSAN

Shuksan Greenschist

Climbers on the summit ridge of Mount Shuksan can appreciate the textures and structures of the Shuksan Greenschist (a major part of the Easton terrane), which makes up the Shuksan massif. The rock is mostly green—naturally—and generally finely layered, with light and dark greens. Commonly, the foliation and the layering are highly crinkled by folding. The minerals chlorite, epidote, and actinolite color the rock green and are its most abundant constituents. The schist, derived from ocean-floor basalt by metamorphism, has a chemical composition just like that of basalt erupted at the mid-ocean ridges, where the ocean-floor plates are

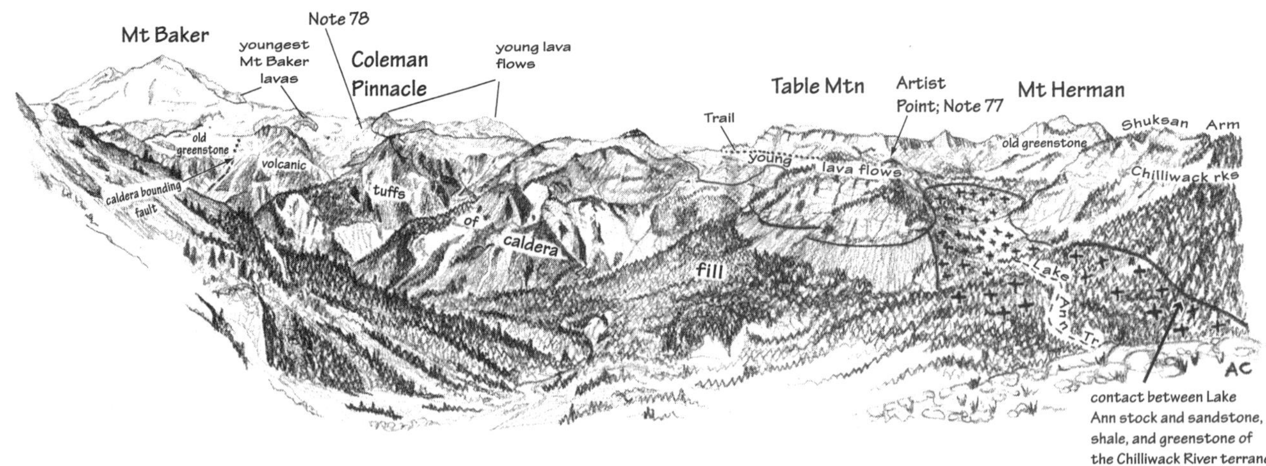

Figure 77. View to the west from the Lake Ann Trail across Swift Creek to Kulshan Caldera and overlying younger volcanic rocks. Crosses indicate location of the Lake Ann stock.

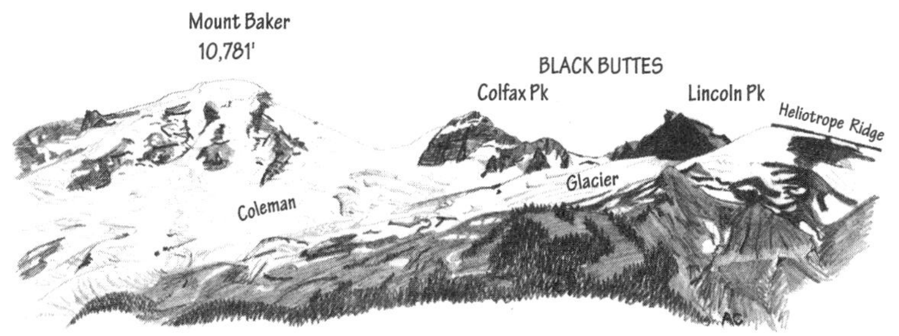

Figure 78. Black Buttes from Glacier Creek Road viewpoint.

generated. The mountain itself is a remnant of a once-continuous thrust plate that overlay the whole region to the west (Figure 16; Geologic Notes 48, 79).

Not all Shuksan Greenschist is green. Some layers are decidedly blue (but these are more obvious on the north side of the mountain than on the regular route up the south side). The blue color is from an unusual, sodium-rich type of amphibole. The crystals are very small and, even with a hand lens, are not easily distinguished. By making artificial rocks with the same minerals in high-pressure cookers, geologists in the lab have found that these blueschists must have formed at about 700° F and at great pressure, corresponding to that found at a depth of about 15 miles in the Earth. Usually, rocks buried this deeply become much hotter than these did and yield a more common amphibole, such as hornblende.

We know from the blue amphibole that the ocean-floor basalt was taken deep into the Earth and brought back up rapidly before it had time to heat up much. Geologists ascribe such a metamorphic history to subduction followed by rapid uplift and removal of the overlying rocks (Chapter 2). Evidently this geologically accelerated spasm in the Earth's crust is not uncommon in the formation of big mountains, because blueschists occur in belts associated with major mountain ranges throughout the world.

81

CANYON CREEK ROAD

Crossing an active landslide

On the Canyon Creek road (USFS Road 31) where it winds around the ridge far above Canyon Creek (about 4 miles or so from the junction with the main highway), look for an abrupt dip in the road as it crosses the break-away fault of a massive landslide. For about one-third mile the road continues smoothly, crossing a creek, then climbing up a step on the far (north) side of the slide. This slide is constantly reactivated by wet weather. No one knows if it will eventually fail catastrophically and plunge into Canyon Creek below, taking the road with it.

GLACIER CREEK ROAD (USFS ROAD 39) AND BEYOND

COLEMAN GLACIER VISTA

Grand views of old terranes capped by a young volcano

Beyond the trailhead for the Coleman Glacier Trail, Glacier Creek Road leads to a high vista point (11 miles from State Route 542) and a superb view of Mount Baker, not to speak of the rocks to contemplate. Actually, openings along the road just before the vista point may provide better views. The route up Glacier Creek is in sandstone and argillite of the Nooksack terrane. Beds of the mostly black to gray rocks are visible in several places where the road crosses creeks. These rocks are at the bottom of the thick pile of faulted thrust plates making up the Western Domain (Chapter 3; see Plate 2). Just above the vista point to the west, the black, cliffy promontory of rock is an erosional remnant of another faulted thrust slab. This slab is a slice of the very old Yellow Aster Complex of the Bell Pass Mélange, here resting directly on the Nooksack terrane, where geologists and readers of this book would normally expect to find rocks of the Chilliwack River terrane.

Of course, the young volcano, Mount Baker, really steals the show here. The dark peaks sticking up above the Coleman Glacier, on the west ridge of Mount Baker, are the Black Buttes, remnants of an older volcano that erupted about 500,000 to 300,000 years ago (Figure 78; Geologic Note 55).

From this vantage, morainal ridges are visible, extending downslope from the Coleman Glacier between cascading streams and ending in the forest below. Far across Glacier Creek to the east, the gravelly slopes of the Chromatic Moraine perch high above the present Glacier Creek. This lateral moraine was built when the Roosevelt Glacier was much bigger and thicker, probably extending much of the way down the present Glacier Creek valley. Like a bathtub ring, the ramplike top of the moraine (marked by trees) remains, revealing the height of the glacier's surface when the moraine was deposited (Chapter 7; Figure 37).

83

COLEMAN GLACIER TRAIL

A close-up view of the glacier and its morainal cloak

Visitors seeking great views and an Ice Age atmosphere will find it by hiking the Coleman Glacier Trail to the edge of the Coleman Glacier. The trail winds in and out of small, steep gorges with cascading creeks. Ridges of Nooksack sandstone and shale are mantled with morainal debris. In the meadows and dwindling forest, a side trail leads east (left) to an overlook above the Coleman Glacier (about 3 miles from the road). Mountaineers practice their glacier-rescue techniques on the crevassed ice below. A glacier's crevasses advertise its motion. The lower ice of the glacier is under great pressure from the overlying ice and deforms under the weight—oozing slowly downhill. The upper part, however, is brittle, and as the lower part creeps around corners or over bumps, the upper part breaks and cracks. Crevasses rarely are deeper than 125 feet, the depth at which ice becomes so plastic and fast-flowing that cracks close up as soon as they are formed.

Crevasses are troublesome to mountaineers hiking on a glacier because they must often cross the icy fissures on unstable snow bridges. In winter, even very wide crevasses can be bridged by drifting snow. In summer these snow bridges commonly thin out from melting on their undersides, where the damage is hidden from climbers. Even experienced climbers may not know that they are about to step through a thin bridge of snow. Fortunately, at least on the lower end of a glacier, most crevasses are revealed during the summer months of melting.

The Coleman Glacier is covered with much dirt and rock debris. This thin carpet of debris accumulates as hills of moraine where the ice melts at the snout of the creeping glacier.

SKYLINE DIVIDE

Proof of the ocean

The rocks along the Skyline Divide Trail seem dark and drab. Mostly, they are black argillite, occasionally stained brown by iron oxides (limonite). Across the valley of the North Fork of the Nooksack, where the same rocks are spectacularly displayed on Excelsior Ridge, the sedimentary beds are more readily seen (see also Plate 4A). Farther up the climbers' trail, which winds along the ridgetop to about 6,300 feet elevation (3.5 miles, 2,000 feet of climbing from the road), outcrops containing fossil clams show these rocks of the Nooksack terrane to be truly ocean-born (Chapter 3). These clams, commonly a variety called *Buchia,* lived in the ocean some 140 million years ago (Figure 79). More easily found than the clams are cylindrical holes in the black argillite, where belemnite remains have mostly dissolved away. Tubular belemnites were long-ago cousins to the nautilus (Figure 17). Ardent fossil enthusiasts can continue up to Chowder Ridge (about 6,800 feet), where clams abound.

Figure 79. A fossil clam (Buchia), typical of the Nooksack Formation.

SWAMP CREEK–TWIN LAKES ROAD (USFS 3065) AND BEYOND

TWIN LAKES ROAD

Cutting the deck

The road to Twin Lakes is a private mine road, a tentacle of civilization into the Mount Baker Wilderness. Many years the road is closed by mud slides and snow-avalanche

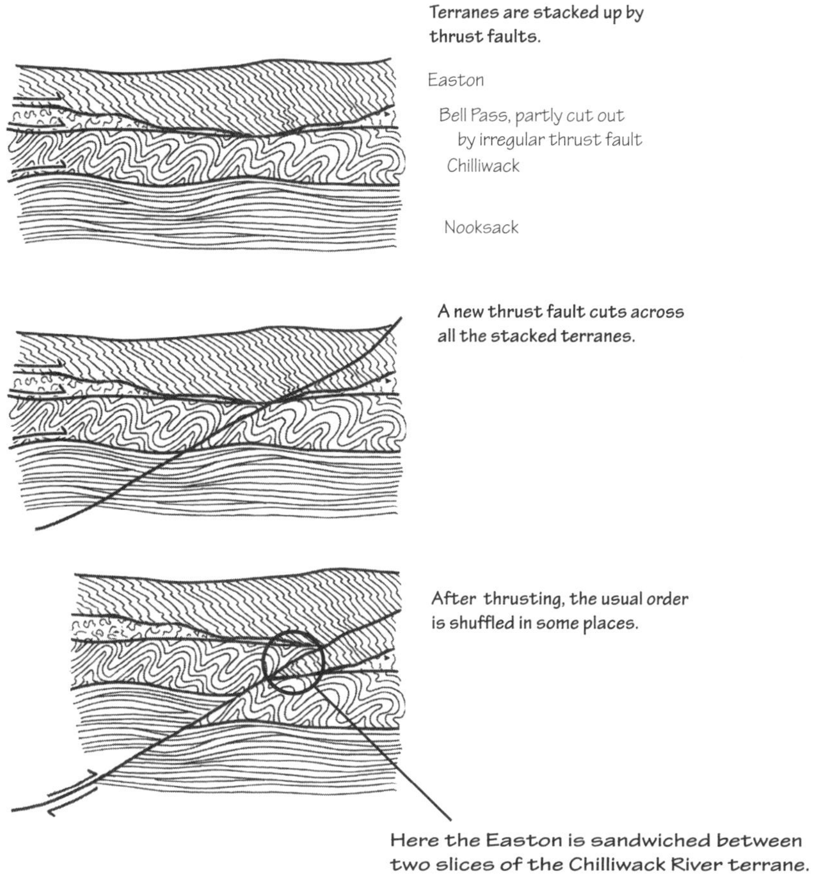
Terranes are stacked up by thrust faults.
Easton
Bell Pass, partly cut out by irregular thrust fault
Chilliwack
Nooksack
A new thrust fault cuts across all the stacked terranes.
After thrusting, the usual order is shuffled in some places.
Here the Easton is sandwiched between two slices of the Chilliwack River terrane.

Figure 80. A last whack of thrusting shuffles the usual pile of thrust plates.

debris (Geologic Note 63), but most summers the mine owners or the friends of Twin Lakes open the road—sort of. All but four-wheel-drive vehicles generally must be parked near the Forest Service trail to Gold Run Pass and Tomyhoi Lake (4.6 miles from the Mount Baker Highway). From the road, but beyond the Gold Run Pass Trail junction, look upslope to cliffs of dark phyllite, a layer of Darrington Phyllite (Easton terrane) sandwiched between thrust plates of volcanic rocks of the Chilliwack River terrane. Throughout most of the Western Domain, the Chilliwack River terrane lies below the Easton terrane (Figure 15), but not both above and below, as is the case here. Such a glaring exception to the structural order might destroy the faith of any geologic acolyte. But as shown in Figure 80, the anomalous stacking is simply the result of subsequent thrust-faulting that cut across the deck of previously stacked terranes.

The road switchbacks up through the Chilliwack volcanic rocks, but just across Swamp Creek to the south, the shoulder of Goat Mountain is all Darrington Phyllite. More views of this complex geology may be enjoyed from Winchester Mountain (Geologic Note 87).

86

TWIN LAKES

Ancient volcanic rocks and a glacial pass

Twin Lakes are glacial tarns nestled in a U-shaped glacier-carved pass. The brown and rubbly rocks rising up from the evenness of the pass are volcanic rocks of the Chilliwack River terrane, and with a bit of looking, visitors can find fragments or pods of light gray, roughened limestone. Light-colored outcrops on the hillside, which contrast with the darker volcanic rocks, may be the same material. Some of the pods in the volcanic rocks have fossils (Geologic Note 1). The volcanic rocks themselves are so altered by metamorphism that on close

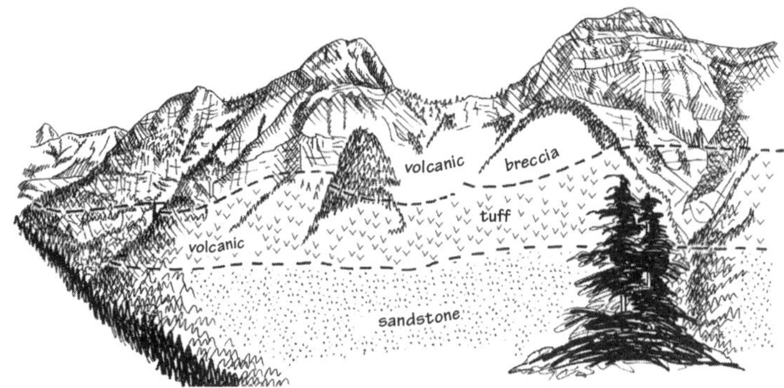

Figure 81. Looking east at Big Bosom Buttes from Lone Jack Mine. Volcanic breccia, forming rugged peaks, overlies volcanic tuff which in turn overlies older, river-deposited sandstone.

approach they are rather nondescript and unattractive even to a geologist.

87

WINCHESTER MOUNTAIN

View of geologic confusion

The trail to the lookout on Winchester Mountain climbs through volcanic flows and breccias of the Chilliwack River terrane. The summit offers a panoramic view of a collage of tectonic terranes. From Yellow Aster Meadows on the west to Mount Larrabee on the north, the rocks belong to the Bell Pass Mélange (mostly the Yellow Aster Complex) and Chilliwack River terrane. These rocks crop out in a confusing pattern, suggesting a great shuffling of the usual thrust plates, further scrambled by steep, cross-cutting faults (see Plate 2). On the south, rocks of the Easton terrane dominate: Goat Mountain in the near view is held up by Darrington Phyllite; Mount Shuksan beyond is of Shuksan Greenschist.

In 1952, Peter Misch outlined the rock pattern here. More of the complex details of this scene were painstakingly mapped by mountain-climbing Professor Ned Brown and his students at Western Washington University. Only by looking at almost every outcrop can geologists begin to make much sense of rocks such as these.

LONE JACK MINE AND A VIEW OF BIG BOSOM BUTTES

Gold and an ancient caldera

Miners developed the Lone Jack Mine in the 1890s, and by 1924, when the mine closed, they had removed several hundred tons of ore. Over the years, avalanches and fire have destroyed several concentrating mills and other buildings on this steep valley wall. Quartz veins with gold in the black phyllite have attracted miners right up to the present day. The source of the gold-bearing veins may be granitic root plutons of the Cascade Arc, which are exposed in Silesia Creek, below to the east.

Across the valley to the east, Big Bosom Buttes reveal the remains of a volcanic caldera that erupted 25–30 million years ago (Oligocene), during the early years of the Cascade Volcanic Arc (Figure 81; Chapter 6). The volcanic rocks overlie older stream deposits of sandstone. These older rocks are remnants of extensional deposits (Chapter 5) that filled fault-bounded valleys before the birth of the volcanic arc.

HIGH PASS AND THE PLEIADES

Shuffled rocks on a grand scale

The old mine trail to Mount Larrabee affords a dramatic view of smashed and mixed rocks of the Western Domain. Where the trail climbs from High Pass up through volcanic and sedimentary rocks of the Chilliwack River terrane, look northward to the steep slopes of Mount Larrabee to see slivers and lenses of light-colored hard rocks usually associated with the Bell Pass Mélange, but mixed here by faulting into sedimentary rocks of the Chilliwack River terrane (Figure 82). The metamorphic rocks of gneiss and schist, as well as chips of ultramafic rocks from the Earth's mantle, have been thoroughly mixed with the sedimentary rocks of the Chilliwack River terrane, probably through subsequent faulting of the thrust-stacked terranes of the Western Domain (Geologic Note 85). These mixed rocks can also be

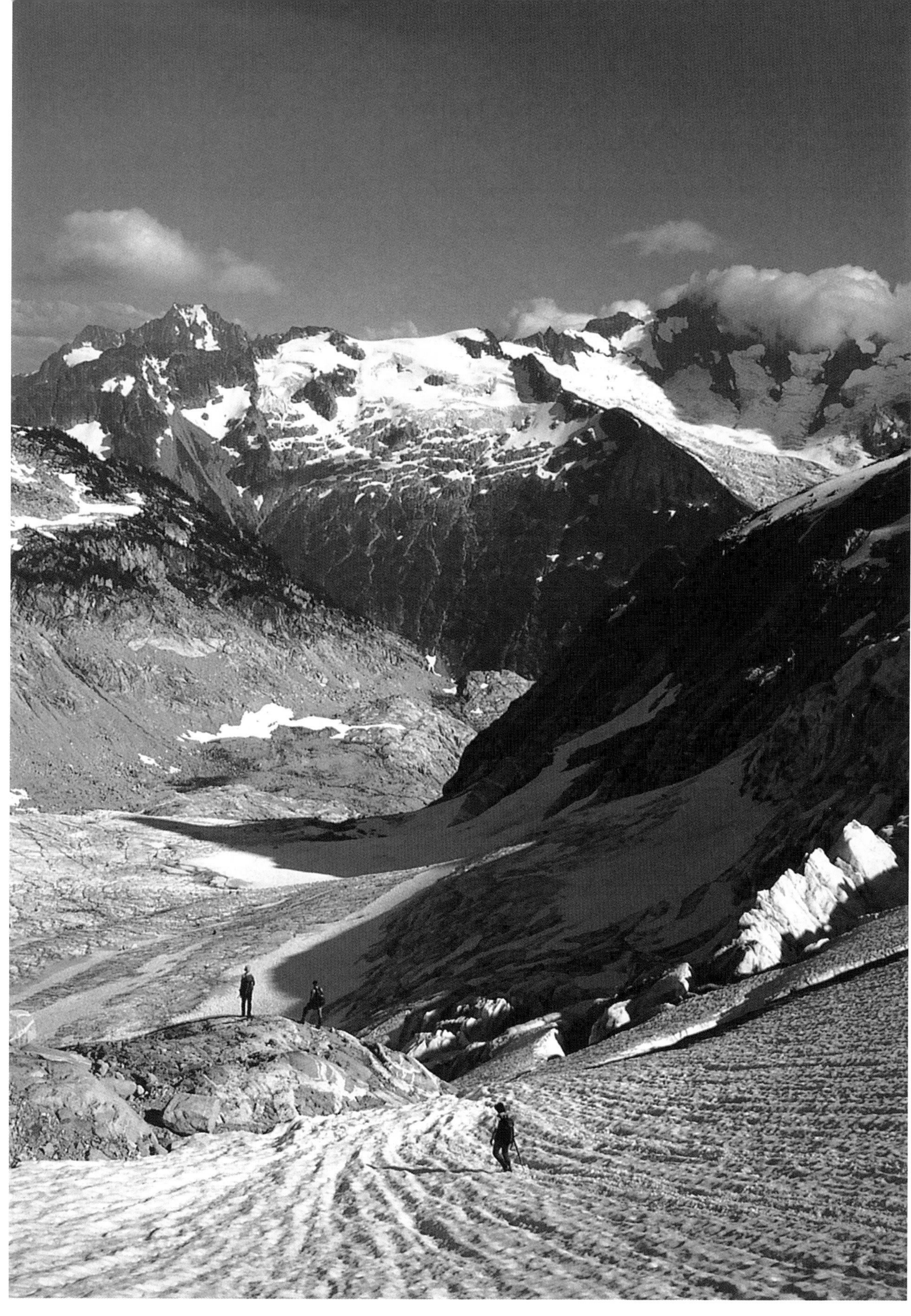

Plate 1. *In the right foreground, where it descends steeply, North Klawatti Glacier breaks into a heavily crevassed icefall. Almost all the visible rock is Eldorado Orthogneiss, a 90-million-year-old stitching pluton in the Metamorphic Core Domain. Mount Buckner is on the left skyline and Mount Forbidden is hidden in the clouds.*

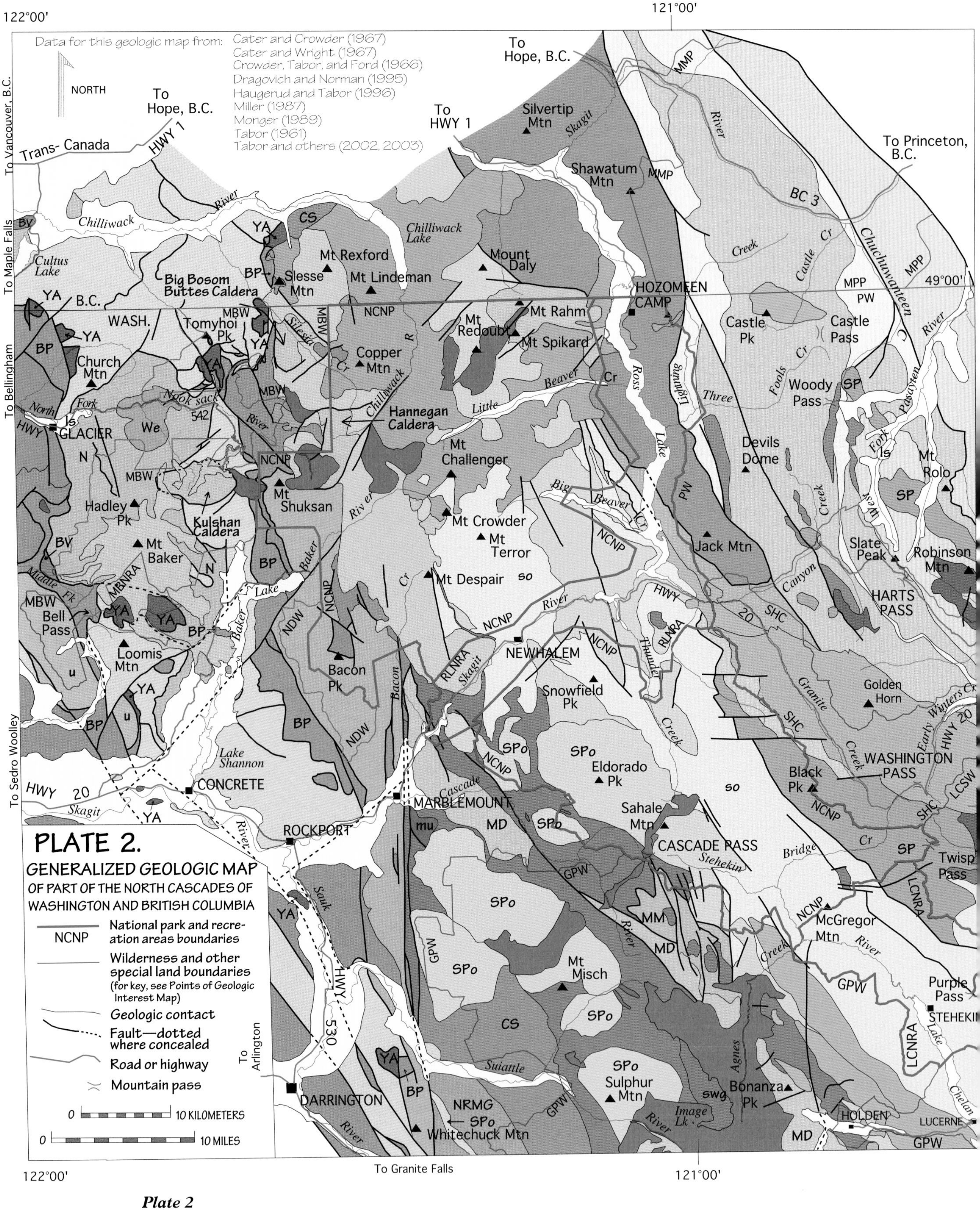

PLATE 2.
GENERALIZED GEOLOGIC MAP
OF PART OF THE NORTH CASCADES OF WASHINGTON AND BRITISH COLUMBIA
NCNP National park and recreation areas boundaries
Wilderness and other special land boundaries (for key, see Points of Geologic Interest Map)
Geologic contact
Fault—dotted where concealed
Road or highway
Mountain pass
0 10 KILOMETERS
0 10 MILES
Data for this geologic map from:
Cater and Crowder (1967)
Cater and Wright (1967)
Crowder, Tabor, and Ford (1966)
Dragovich and Norman (1995)
Haugerud and Tabor (1996)
Miller (1987)
Monger (1989)
Tabor (1961)
Tabor and others (2002, 2003)
NORTH
122°00'
121°00'
49°00'
To Hope, B.C.
To HWY 1
To Princeton, B.C.
To Vancouver, B.C.
To Maple Falls
To Bellingham
To Sedro Woolley
To Arlington
To Granite Falls
Trans- Canada
HWY 1
BC 3
B.C.
WASH.
Chilliwack River
Cultus Lake
Chilliwack Lake
Big Bosom Buttes Caldera
Hannegan Caldera
Kulshan Caldera
Slesse Mtn
Mt Rexford
Mt Lindeman
Mount Daly
Silvertip Mtn
Shawatum Mtn
HOZOMEEN CAMP
Tomyhoi Pk
Church Mtn
Copper Mtn
Mt Redoubt
Mt Rahm
Mt Spickard
Castle Pk
Castle Pass
Woody Pass
Devils Dome
Mt Rolo
Jack Mtn
Slate Peak
Robinson Mtn
HARTS PASS
GLACIER
Mt Shuksan
Mt Challenger
Mt Crowder
Mt Terror
Mt Despair
Hadley Pk
Mt Baker
Bell Pass
Loomis Mtn
Bacon Pk
NEWHALEM
Snowfield Pk
Golden Horn
Eldorado Pk
Sahale Mtn
CASCADE PASS
Black Pk
WASHINGTON PASS
Twisp Pass
McGregor Mtn
Purple Pass
STEHEKIN
CONCRETE
ROCKPORT
MARBLEMOUNT
Lake Shannon
Mt Misch
Sulphur Mtn
Bonanza Pk
Image Lk.
HOLDEN
LUCERNE
DARRINGTON
Whitechuck Mtn
HWY 20
HWY 530
542
Ross Lake
Baker Lake
Skagit River
Sauk River
Suiattle River
Stehekin River
Nooksack River
Chuchuwanteen Creek
Lightning Creek
Three Fools Creek
Canyon Creek
Granite Creek
Thunder Creek
Bridge Creek
Agnes Creek
Early Winters Cr
Lake Chelan
Pasayten River
NCNP
RLNRA
LCNRA
MBNRA
LCSW
MMP
MPP
PW
SP
SHC
Is
CS
YA
BY
BP
BV
MBW
We
N
u
NDW
so
SPo
MD
mu
GPW
MM
NRMG
swg

Plate 2

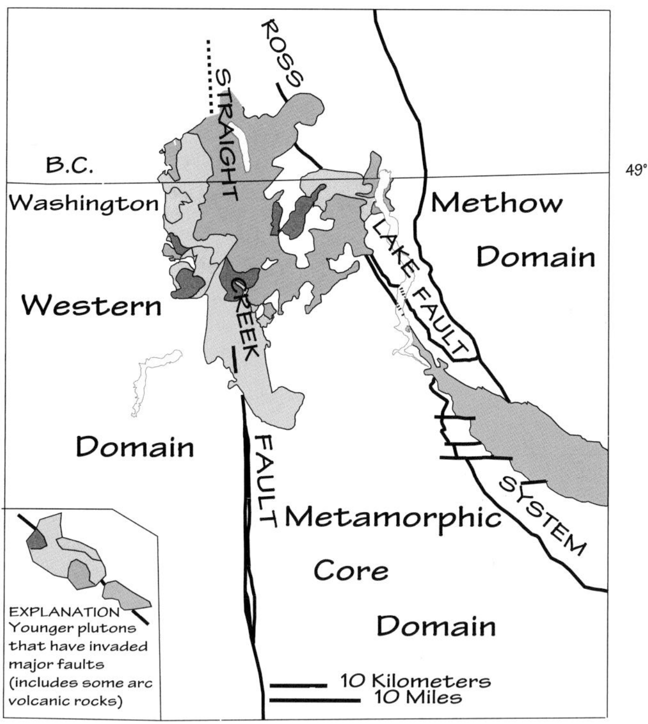

Area of Geologic Map showing major faults, domains, and a few young plutons that obscure the faults.

EXPLANATION

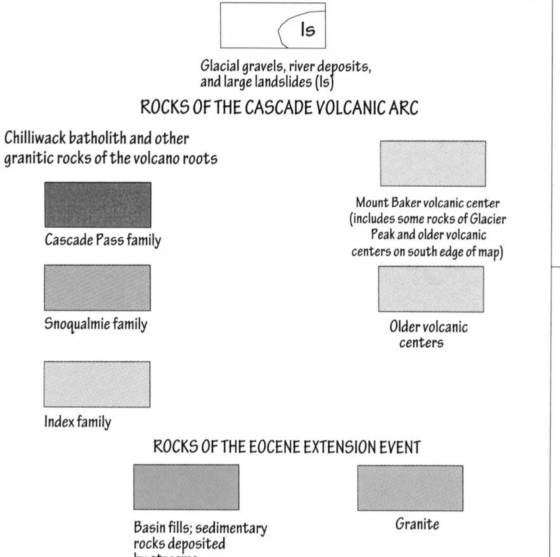

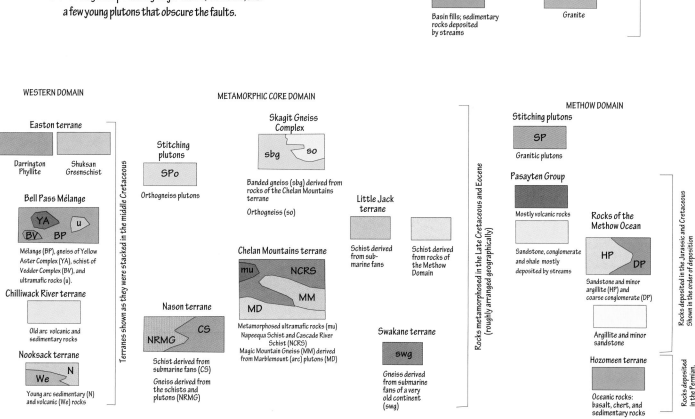

WESTERN DOMAIN

Easton terrane

Darrington Phyllite

Shuksan Greenschist

Bell Pass Mélange

YA u BV BP

Mélange (BP), gneiss of Yellow Aster Complex (YA), schist of Vedder Complex (BV), and ultramafic rocks (u).

Chilliwack River terrane

Old arc volcanic and sedimentary rocks

Nooksack terrane

We N

Young arc sedimentary (N) and volcanic (We) rocks

Terranes shown as they were stacked in the middle Cretaceous

METAMORPHIC CORE DOMAIN

Stitching plutons

SPo

Orthogneiss plutons

Skagit Gneiss Complex

sbg so

Banded gneiss (sbg) derived from rocks of the Chelan Mountains terrane

Orthogneiss (so)

Little Jack terrane

Schist derived from sub-marine fans

Schist derived from rocks of the Methow Domain

Chelan Mountains terrane

mu NCRS MM MD

Metamorphosed ultramafic rocks (mu)
Napeequa Schist and Cascade River Schist (NCRS)
Magic Mountain Gneiss (MM) derived from Marblemount (arc) plutons (MD)

Nason terrane

NRMG CS

Schist derived from submarine fans (CS)

Gneiss derived from the schists and plutons (NRMG)

Swakane terrane

swg

Gneiss derived from submarine fans of a very old continent (swg)

Rocks metamorphosed in the Late Cretaceous and Eocene (roughly arranged geographically)

METHOW DOMAIN

Stitching plutons

SP

Granitic plutons

Pasayten Group

Mostly volcanic rocks

Sandstone, conglomerate and shale mostly deposited by streams

Rocks of the Methow Ocean

HP DP

Sandstone and minor argillite (HP) and coarse conglomerate (DP)

Argillite and minor sandstone

Rocks deposited in the Jurassic and Cretaceous Shown in the order of deposition

Hozomeen terrane

Oceanic rocks: basalt, chert, and sedimentary rocks

Rocks deposited in the Permian,

Plate 3A. *American Border Peak (left) and Yellow Aster Butte (right) are both carved from volcanic rocks of the Chilliwack River terrane. Yellow Aster Meadows (foreground) are held up by large slab of Yellow Aster granitic gneiss in the Bell Pass Mélange (Geologic Note 90). The gray rock to the right of American Border Peak is a small arc-root pluton. Rocks in the contact zone around it weather rusty brown.*

Plate 3B. *Late summer snow highlights beds of sandstone from the Methow Ocean (Harts Pass Formation). Looking south along the Cascade Crest from Mount Winthrop.*

Plate 3C. *Remnants of the ancient Black Buttes volcano rise up above the Deming Glacier. A younger member of the Cascade Volcanic Arc, the volcano erupted about 500,000 years ago.*

4A.

Plate 4A. *Mount Baker rises behind beds of argillite in the Nooksack Formation on Skyline Divide (Geologic Note 84).*
Plate 4B. *Banded gneiss in the Skagit Gneiss Complex is crosscut by a sharply bounded granite orthogneiss dike. Such dikes are the caulk of the Eocene extensional event. Head of Torrent Creek, west of Sourdough Mountain.*
Plate 4C. *Cascade Peak (left middle distance) is held up by contact-metamorphosed Cascade River Schist. Mixup Mountain (right) is carved from Magic Mountain Gneiss. The aligned snow patches and talus stretching from the foreground to Cache Col (right of Mixup) show the location of the Cache Col Fault, where the crushed rock erodes a little faster to make a swale (Geologic Note 72). Eldorado Peak, underlain by the Eldorado Orthogneiss, is on the skyline at far left.*

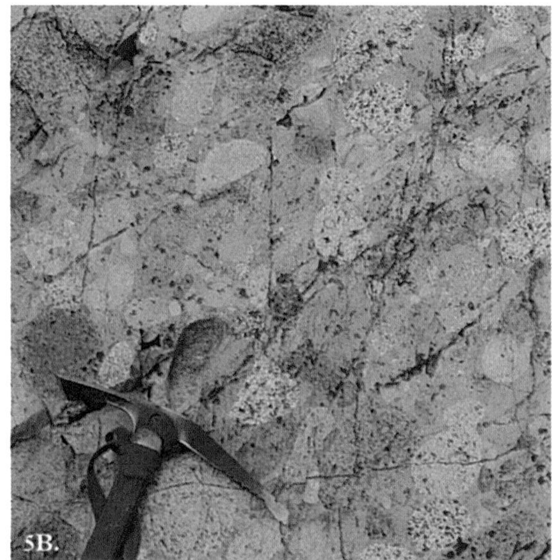

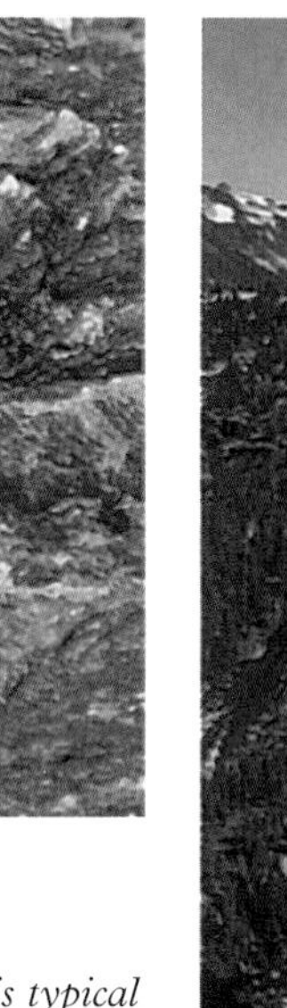

Plate 5A. *A blocky joint pattern and light color is typical of granitic rocks such as the tonalite and granodiorite of the Chilliwack batholith (here being traversed by geologists) on west ridge of Mount Challenger (photo by John Harbuck).*

Plate 5B. *Conglomerate of the Methow Ocean near Woody Pass. The cobbles of granitic rock (white with black speckles) may have come from mountains eroded in Mexico (Geologic Note 150).*

Plate 5C. *Fragments of greenstone, sandstone, and other rocks in disrupted argillite. Bell Pass Mélange below the Shuksan Thrust Fault, south side of Suiattle Mountain.*

Plate 5D. *Found Lakes were carved by glaciers from the granodiorite of a stitching pluton in the Cascade Metamorphic Core. The foreground lake has no glacial silt in the water; the far one does.*

LOCATION MAP FOR POINTS OF GEOLOGIC INTEREST MAPS

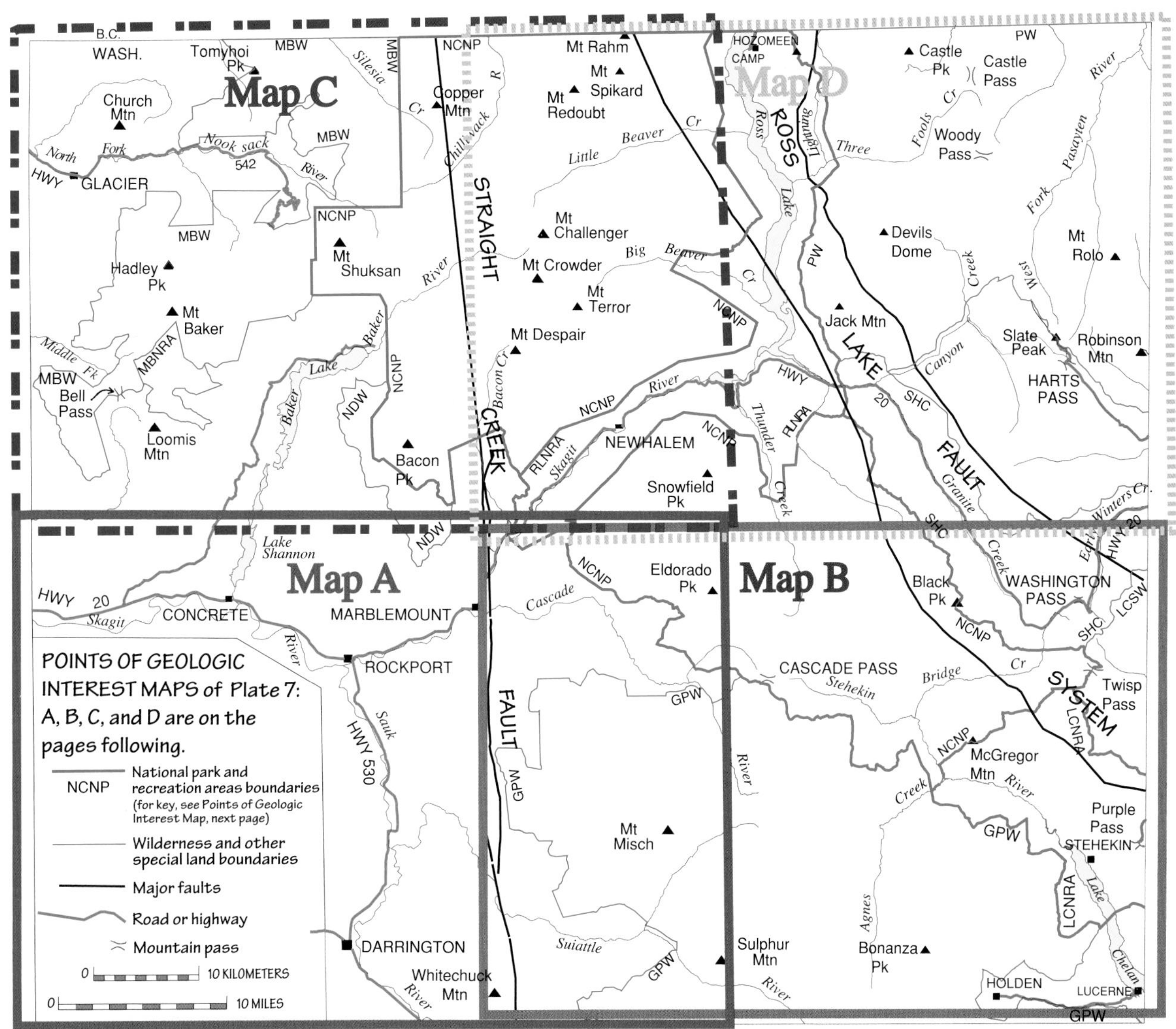

Technical note: The shaded-relief base for Maps A, B, C, and D was generated by computer from a grid of elevation points. These elevation points were mostly derived from 1:24,000-scale contour maps. Faint E-W furrows on some slopes are artifacts of bad data.

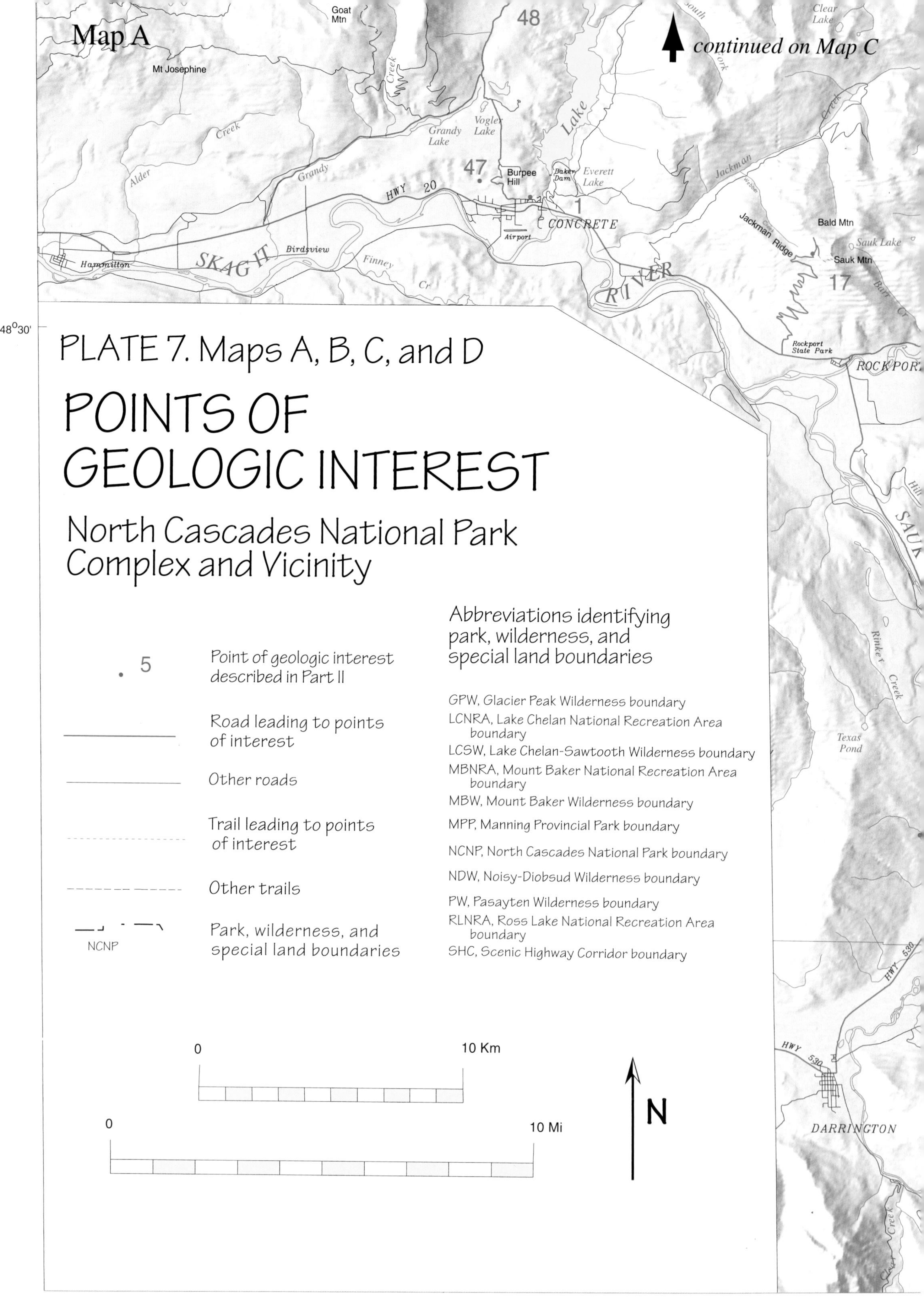
Map A
continued on Map C
Goat Mtn
48
Clear Lake
Mt Josephine
Creek
Grandy Lake
Vogler Lake
Lake
Alder
Creek
Grandy
47
Burpee Hill
Baker Dam
Everett Lake
Jackman
HWY 20
1
CONCRETE
Airport
Jackman Ridge
Bald Mtn
Sauk Lake
Sauk Mtn
17
Hamilton
Birdsview
SKAGIT
Finney
Cr
RIVER
Rockport State Park
ROCKPORT
48°30'
PLATE 7. Maps A, B, C, and D
POINTS OF GEOLOGIC INTEREST
North Cascades National Park Complex and Vicinity
. 5
Point of geologic interest described in Part II
Road leading to points of interest
Other roads
Trail leading to points of interest
Other trails
NCNP
Park, wilderness, and special land boundaries
Abbreviations identifying park, wilderness, and special land boundaries
GPW, Glacier Peak Wilderness boundary
LCNRA, Lake Chelan National Recreation Area boundary
LCSW, Lake Chelan-Sawtooth Wilderness boundary
MBNRA, Mount Baker National Recreation Area boundary
MBW, Mount Baker Wilderness boundary
MPP, Manning Provincial Park boundary
NCNP, North Cascades National Park boundary
NDW, Noisy-Diobsud Wilderness boundary
PW, Pasayten Wilderness boundary
RLNRA, Ross Lake National Recreation Area boundary
SHC, Scenic Highway Corridor boundary
0
10 Km
0
10 Mi
N
SAUK
Rinker Creek
Texas Pond
HWY 530
HWY 530
DARRINGTON
Creek

continued on Map C
continued on Map B
RLNRA
Lake Stillwell
Helen Cr
Olson Lake
Helen Buttes
Olson
Creek
Rocky
Corkindale
Creek
Patent Cr
Bald Lake
Sauk Lake
MARBLEMOUNT
SKAGIT
Copper
Alma
NATIONAL
PARK
FOREST
Monogram Lake
Haystack Cr
NORTH
65
66
Lookout Mtn
Taylor Cr
Day Cr
Lookout Cr
Marble Creek
Dorado Needle
Eldorado Peak
2
CASCADE 61
Sibley Creek
The Triad
Roush Cr
67
Hidden Lake
Hidden Lake Peaks
Boulder
Irene
Creek
Jordan
Creek
Barr Cr
ROCKPORT
O'Brien Cr
Illabot
Creek
Granite Lakes
RIVER
Mineral Park
GPW
Jordan Lakes
Hilt
Creek
Creek
Iron Cr
Slide Lake
Found
Found Lakes
Kindy
Creek
62
SAUK
Creek
Bluebell Cr
Arrow
Illabot Peaks
Cyclone Lake
Snowking Mtn
White
Suiattle Mtn
Creek
Rinker Creek
Grade
Creek
Cr
Jug Lake
GLACIER
Mt Chaval
Bluff Lake
GPW
Crater Lake
Big
Creek
Mt Buckindy
Mt Misch
Goat
SUIATTLE
Tenas
Boulder Lake
Creek
Creek
Creek
107
Creek
Horse
PEAK
Horse Lake
Bachelor
Huckleberry
Mountain
109
Green Mtn
Downey Lake
Downey
HWY 530
Prairie Mtn
Creek
All
108
Downey Mtn
Sulphur
Dan
110
Sulpher Mtn Lake
DARRINGTON
Creek
Cr
Straight
RIVER
Circle
Lime
111
Sulphur Mtn
112
Gold
Mountain
Decline
Creek
Black
Cr
Circle Pk
Box Mtn
GPW
Conn
Creek
Creek
White Chuck Mtn
Indigo Lake
Lime
Rivord Lake
RIVER
Rat Trap Pass
Crystal
Cr
Crystal Lake
Creek
Ridge
Milk Lakes
Milk Cr
Meadow Mtn
Emerald Lake

Map B
continued on Map D
continued on Map A
Big Devil Peak
SKAGIT
RIVER
NATIONAL
NATIONAL
PARK
FOREST
NORTH
CASCADES
CASCADE
Lookout Mtn
Monogram Lake
Primus Peak
Tricouni Peak
Austera Towers
Dorado Needle
Klawatti Peak
Eldorado Peak
The Triad
Mt Torment
Forbidden Peak
Boston Glacier
Hidden Lake
Hidden Lake Peaks
SKAGIT CO
CHELAN CO
Buckner Mtn
Sahale Mtn
Horseshoe Basin
Cascade Peak
Cascade Pass
Booker Mtn
Johannesburg Mtn
Mineral Park
The Triplets
Mixup Mtn
Magic Mtn
Pelton Peak
GPW
Trapper Mtn
Mt Formidable
Spider Mtn
Snowking Mtn
Le Conte Mtn
Rimrock
Mt Chaval
Sentinel Peak
Old Guard Pk
Mt Buckindy
Mt Misch
GLACIER
PEAK
WILDERNESS
Spire Pt
Dome Pk
Sinister P
SKAGIT CO
SNOHOMISH CO
Huckleberry Mountain
Green Mtn
Downey Mtn
Ross Pass
SUIATTLE
RIVER
Sulphur Mtn
Totem Pass
Box Mtn
Circle Pk
Miners Ridge
White Chuck Mtn
Rat Trap Pass
Lime Ridge
Meadow Mtn
Grassy Point
2
24
25
61
62
63
64
65
66
67
68
69
70
71
72
108
109
110
111
112
113
114
116
127
128
129

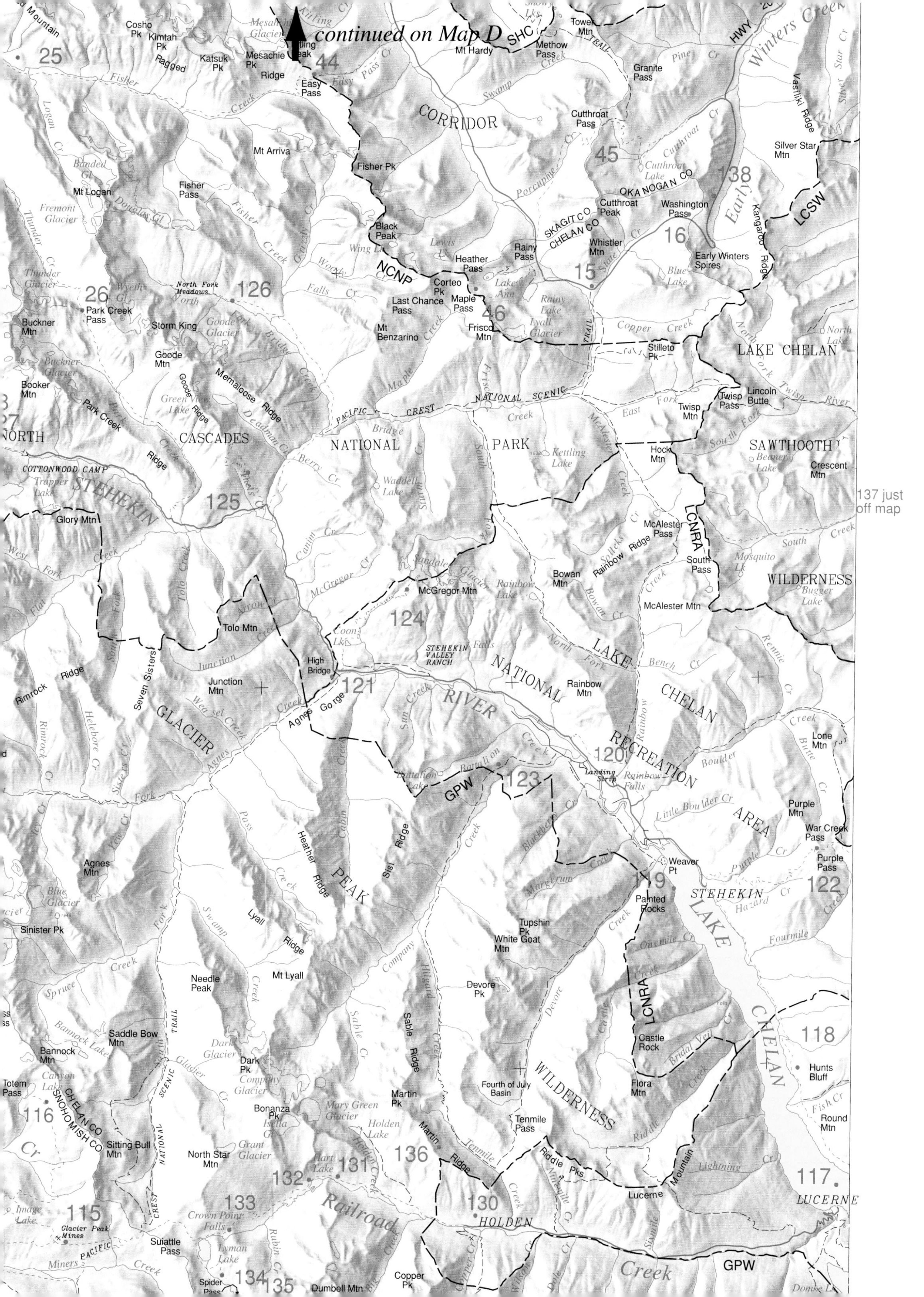

continued on Map D
Cosho Pk
Kimtah Pk
Ragged
Katsuk Pk
Mesachie Pk
Ridge
Easy Pass
Mt Hardy
Methow Pass
Tower Mtn
Granite Pass
Winters Creek
HWY
SHC
CORRIDOR
Cutthroat Pass
Silver Star Mtn
Vasiliki Ridge
Mt Arriva
Fisher Pk
Mt Logan
Fisher Pass
OKANOGAN CO
Cutthroat Peak
Washington Pass
Early Winters Spires
Kangaroo Ridge
LCSW
SKAGIT CO
CHELAN CO
Whistler Mtn
Rainy Pass
Black Peak
Heather Pass
NCNP
Corteo Pk
Maple Pass
Last Chance Pass
Mt Benzarino
Frisco Mtn
North Fork Meadows
Park Creek Pass
Buckner Mtn
Storm King
Goode Mtn
Booker Mtn
Memaloose Ridge
Goode Ridge
Park Creek Ridge
Stilleto Pk
LAKE CHELAN
Twisp Pass
Lincoln Butte
Twisp Mtn
PACIFIC CREST NATIONAL SCENIC TRAIL
NORTH CASCADES NATIONAL PARK
COTTONWOOD CAMP
STEHEKIN
Hock Mtn
SAWTHOOTH WILDERNESS
Crescent Mtn
137 just off map
Glory Mtn
McAlester Pass
LCNRA
Rainbow Ridge
South Pass
Bowan Mtn
McGregor Mtn
McAlester Mtn
Tolo Mtn
STEHEKIN VALLEY RANCH
High Bridge
Rimrock Ridge
Seven Sisters
Junction Mtn
Agnes Gorge
LAKE CHELAN NATIONAL RECREATION AREA
Rainbow Mtn
Lone Mtn
GPW
Landing Strip
Rainbow Falls
Purple Mtn
War Creek Pass
GLACIER PEAK WILDERNESS
Heather Ridge
Sisi Ridge
Agnes Mtn
Weaver Pt
Purple Pass
Painted Rocks
STEHEKIN
Sinister Pk
Lyall Ridge
Tupshin Pk
White Goat Mtn
Needle Peak
Mt Lyall
Devore Pk
Saddle Bow Mtn
Bannock Mtn
Sable Ridge
Castle Rock
Hunts Bluff
Totem Pass
CHELAN CO
SNOHOMISH CO
Dark Pk
Fourth of July Basin
Martin Pk
Flora Mtn
Bonanza Pk
Tenmile Pass
Round Mtn
Sitting Bull Mtn
North Star Mtn
Martin Ridge
Riddle Pks
Mountain
Lucerne
LUCERNE
Railroad Creek
HOLDEN
Glacier Peak Mines
Crown Point Falls
Suiattle Pass
Spider Pass
Dumbell Mtn
Copper Pk
25
44
45
138
16
15
26
126
46
125
124
121
120
123
119
122
118
116
136
132
131
130
133
115
117
134
135

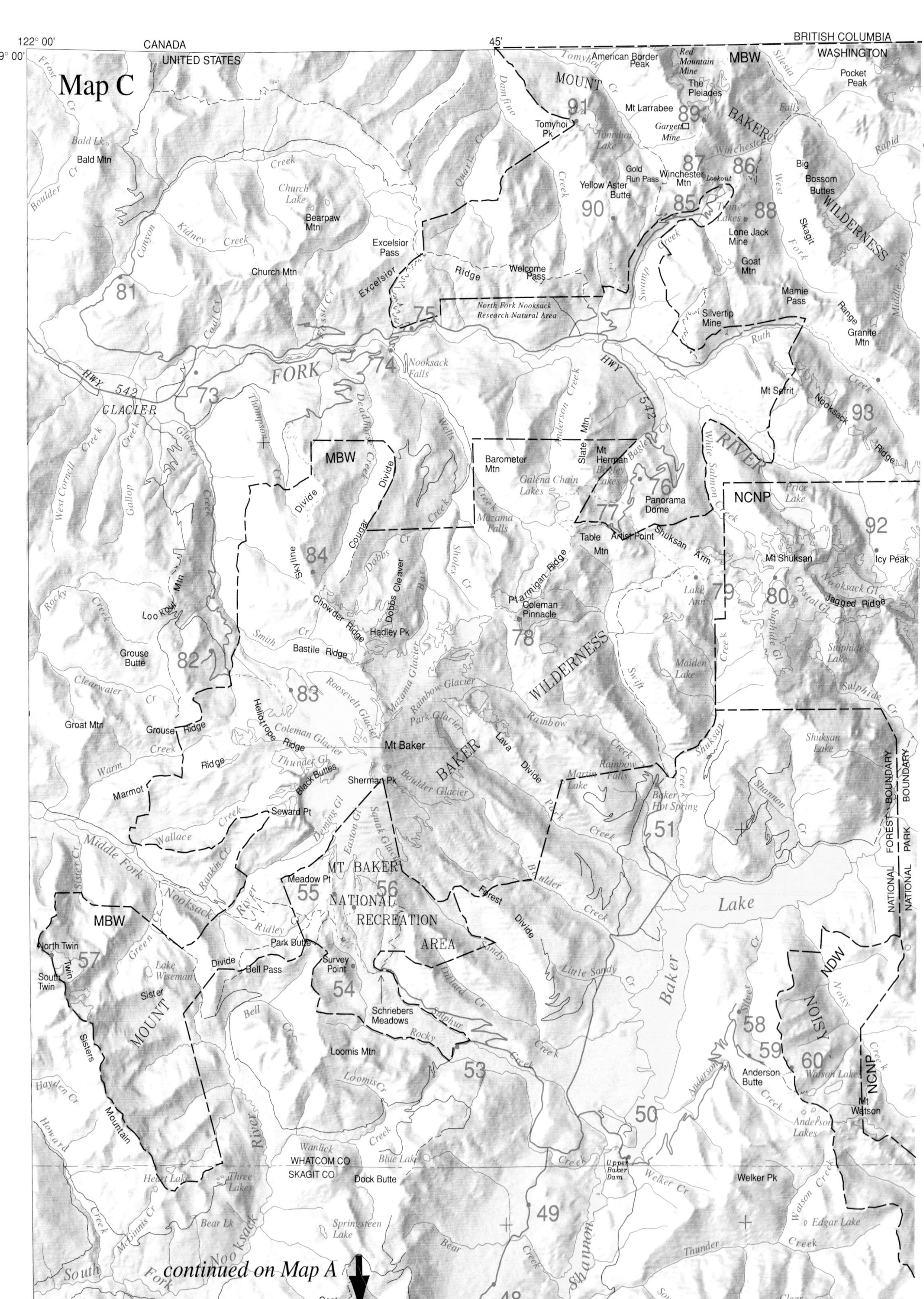

continued on Map A

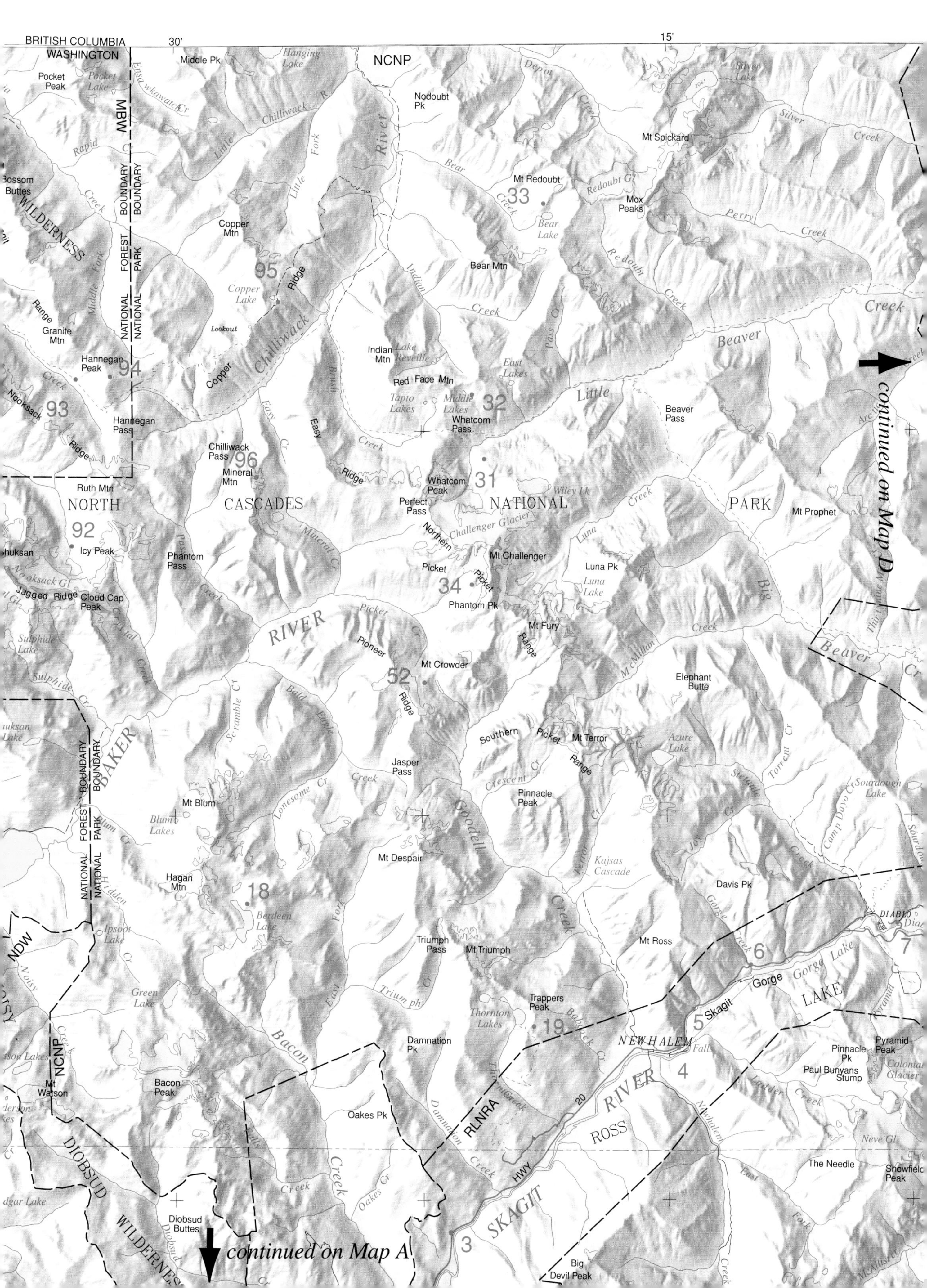
BRITISH COLUMBIA
WASHINGTON
30'
15'
NCNP
Middle Pk
Hanging Lake
Pocket Peak
Pocket Lake
Nodoubt Pk
Depot Creek
Silver Lake
Mt Spickard
Silver Creek
Chilliwack R
Little Fork
River
Rapid Cr
Bossom Buttes
WILDERNESS
Middle Fork
Bear Creek
Mt Redoubt
33
Redoubt Gl
Mox Peaks
Bear Lake
Perry Creek
Redoubt Creek
MBW BOUNDARY
NATIONAL FOREST BOUNDARY
NATIONAL PARK BOUNDARY
Copper Mtn
95
Copper Lake
Ridge
Bear Mtn
Indian Creek
Range
Granite Mtn
Lookout
Chilliwack
Copper
Pass Cr
Beaver
Creek
Hannegan Peak
94
93
Nooksack
Hannegan Pass
Ridge
Indian Mtn
Lake Reveille
Red Face Mtn
East Lakes
Tapto Lakes
Middle Lakes
32
Whatcom Pass
Little
Beaver Pass
Arctic
continued on Map D
Chilliwack Pass
96
Mineral Mtn
Easy Cr
Easy Ridge
Creek
Whatcom Peak
31
Wiley Lk
Ruth Mtn
NORTH CASCADES NATIONAL PARK
Perfect Pass
Mt Prophet
92
Shuksan
Icy Peak
Phantom Pass
Mineral Cr
Northern Picket
Challenger Glacier
Mt Challenger
Luna Pk
Luna Lake
34
Phantom Pk
Nooksack Gl
Jagged Ridge
Cloud Cap Peak
Crystal Creek
Big Creek
Sulphide Lake
RIVER
Picket Cr
Pioneer Ridge
Mt Fury
Range
Mt Crowder
52
McMillan
Beaver Cr
Sulphide Cr
Elephant Butte
Scramble Cr
Bald Eagle
BAKER
Southern Picket Range
Mt Terror
Azure Lake
Jasper Pass
Torrent Cr
Sourdough Lake
Lonesome Cr
Crescent Cr
Pinnacle Peak
Mt Blum
Blum Lakes
Blum Cr
Goodell
Stetattle
Joy Cr
Camp Dayo Cr
Mt Despair
Terror Cr
Kajsas Cascade
Hagan Mtn
18
Hidden
Berdeen Lake
Fork
Creek
Davis Pk
Gorge Creek
DIABLO
NDW
Ipsoot Lake
Triumph Pass
Mt Triumph
Mt Ross
6
7
Gorge Lake
Noisy
Green Lake
East
Triumph
Trappers Peak
Thornton Lakes
19
Babcock Cr
Skagit
5
LAKE
Pyramid
Bacon
Damnation Pk
NEWHALEM
Falls
Pinnacle Pk
Pyramid Peak
Colonial Glacier
Paul Bunyans Stump
NCNP
Mt Watson
Bacon Peak
RLNRA
Thornton Creek
RIVER
4
Ladder Creek
Newhalem
Oakes Pk
20
ROSS
Damnation Creek
Neve Gl
DIOBSUD
Falls
Creek
HWY
The Needle
Snowfield Peak
Oakes Cr
East Fork
Diobsud Buttes
SKAGIT
WILDERNESS
Diobsud
continued on Map A
3
Big Devil Peak
McAllister
Creek

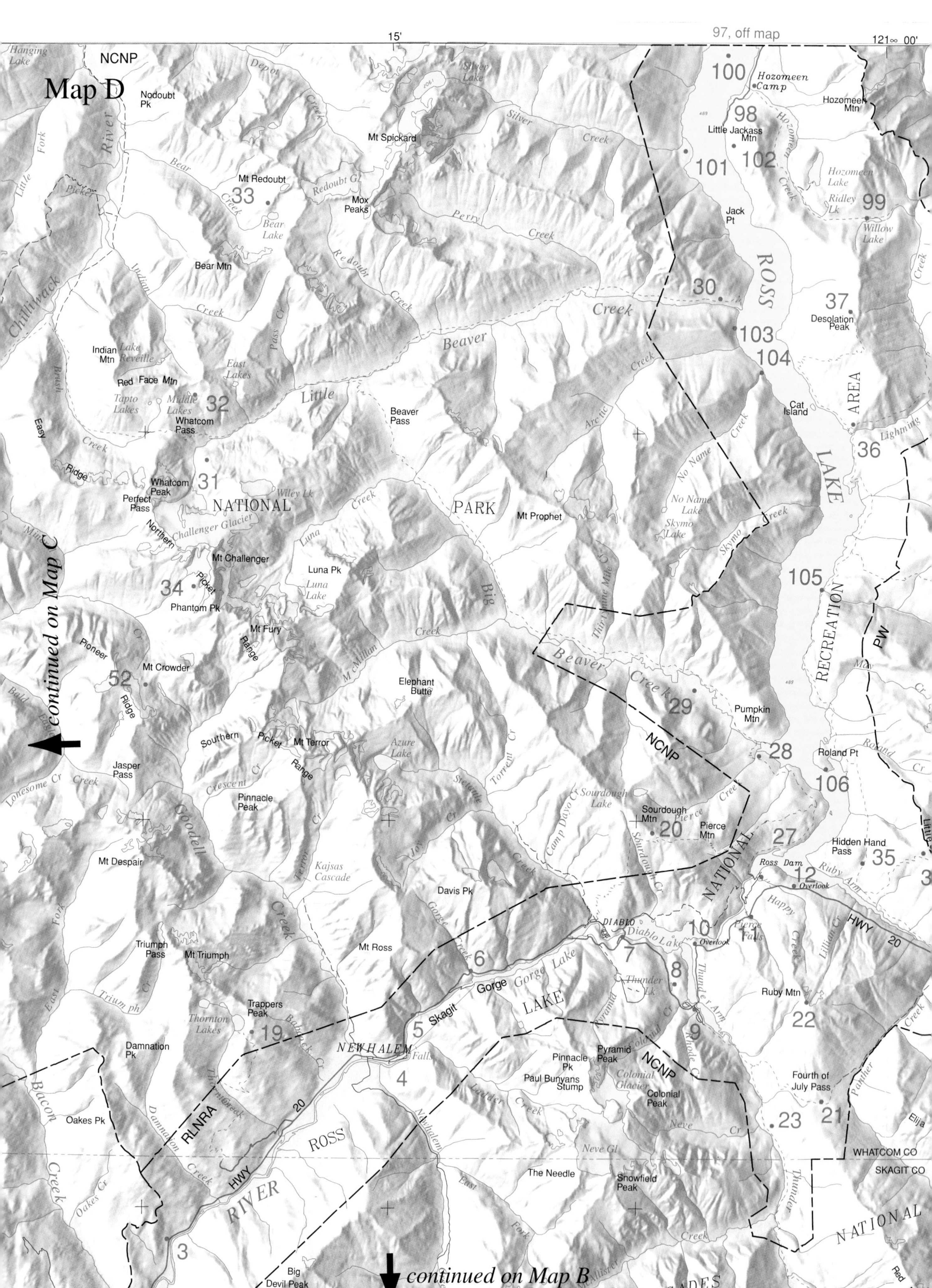
Map D
NCNP
Hanging Lake
Nodoubt Pk
Mt Spickard
Silver Lake
Depot Creek
Silver Creek
Hozomeen Camp
Hozomeen Mtn
Little Jackass Mtn
Hozomeen Creek
Hozomeen Lake
Ridley Lk
Willow Lake
Jack Pt
ROSS LAKE
Mt Redoubt
Redoubt Gl
Mox Peaks
Bear Creek
Bear Lake
Perry Creek
Redoubt Creek
Bear Mtn
Chilliwack
Indian Creek
Pass Cr
Beaver Creek
Little Beaver
Desolation Peak
Cat Island
RECREATION AREA
Lightning
Indian Mtn
Lake Reveille
East Lakes
Red Face Mtn
Tapto Lakes
Middle Lakes
Whatcom Pass
Beaver Pass
Arctic Creek
Easy Ridge
Whatcom Peak
Perfect Pass
NATIONAL PARK
Wiley Lk
Challenger Glacier
Northern Picket Range
Mt Challenger
Mt Prophet
No Name Creek
No Name Lake
Skymo Lake
Skymo Creek
Luna Creek
Luna Pk
Luna Lake
Phantom Pk
Mt Fury
Big Beaver Creek
Thirtynine Mile Cr
continued on Map C
Pioneer Ridge
Mt Crowder
McMillan Creek
Elephant Butte
Pumpkin Mtn
NCNP
PW
Bald Eagle
Southern Picket Range
Mt Terror
Azure Lake
Torrent Cr
Roland Pt
Roland Cr
Jasper Pass
Lonesome Cr
Creek
Crescent Cr
Pinnacle Peak
Stetattle Creek
Sourdough Lake
Camp Davo C
Sourdough Mtn
Pierce Creek
Pierce Mtn
NATIONAL
Hidden Hand Pass
Little
Goodell Creek
Mt Despair
Kajsas Cascade
Terror Cr
Joy Cr
Davis Pk
Sourdough Cr
Ross Dam
Overlook
Ruby Arm
Happy Creek
Lillian Cr
HWY 20
DIABLO
Diablo Lake
Overlook
Pierce Falls
Gorge Creek
Triumph Pass
Mt Triumph
Mt Ross
Gorge Lake
Thunder Lk
Thunder Arm
Ruby Mtn
Triumph Cr
East Fork
Trappers Peak
Skagit
Gorge
LAKE
Pyramid Cr
Colonial Cr
Thornton Lakes
Babcock Cr
Damnation Pk
NEWHALEM
Falls
Pinnacle Pk
Pyramid Peak
NCNP
Rhode Cr
Fourth of July Pass
Panther Creek
Bacon Creek
Paul Bunyans Stump
Colonial Glacier
Colonial Peak
Oakes Pk
Thornton Creek
Damnation Creek
Ladder Creek
Newhalem Creek
Neve Cr
Elija
RLNRA
ROSS
Neve Gl
WHATCOM CO
SKAGIT CO
HWY
RIVER
The Needle
Snowfield Peak
East Fork
Thunder Creek
Oakes Cr
NATIONAL
Big Devil Peak
McAllister Creek
continued on Map B
CASCADES
Red
15'
97, off map
121∞ 00'
100
98
101
102
99
33
30
37
103
104
32
36
31
34
105
52
29
28
106
20
27
35
12
10
11
7
6
8
5
22
19
9
4
23
21
3

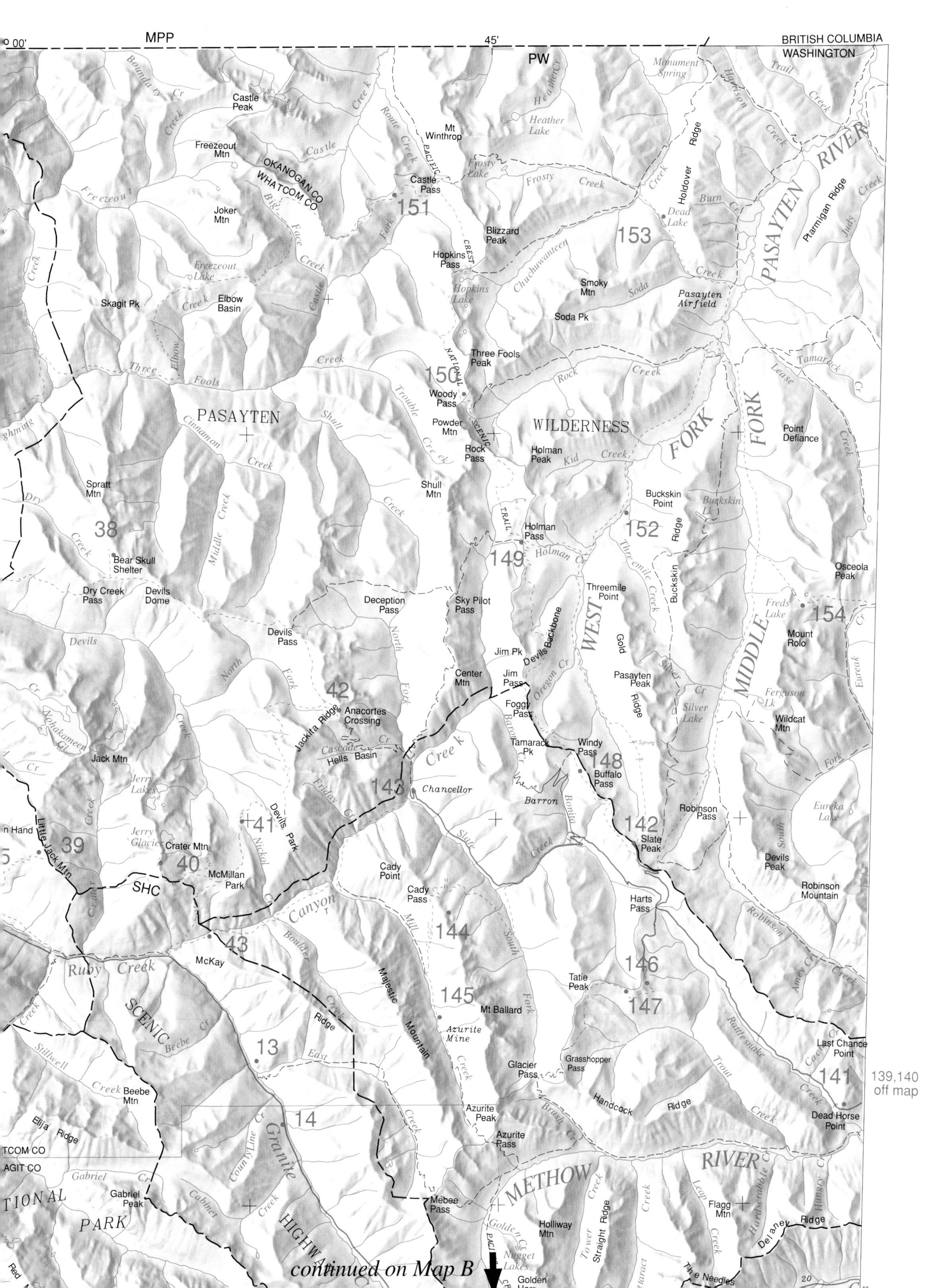

continued on Map B

Plate 8A. *Geologist measuring the inclination of red beds in volcanic rocks of the Pasayten Group, Methow Domain, near Last Chance Point (Geologic Note 140).*

Plate 8B. *Geologist looking at fossilized tree trunk in sandstone deposited by streams during the Eocene extensional event. Chuckanut Formation on Bacon Peak (Geologic Note 60).*

Plate 8C. *A geologist rests on Mount Sefrit (Geologic Note 93) with a view of Mount Shuksan, which is carved from Shuksan Greenschist of the Western Domain.*

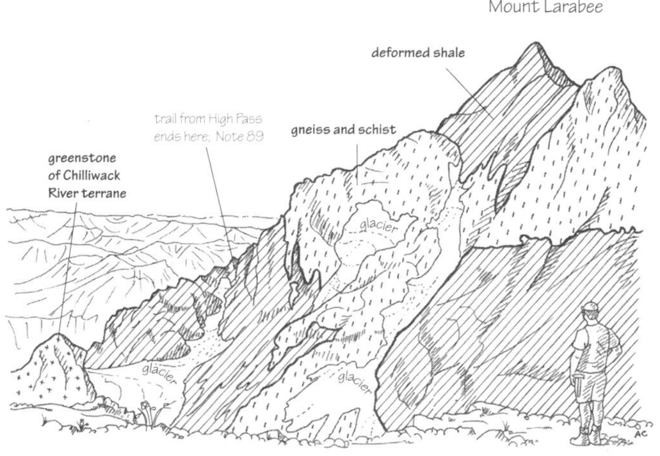

Figure 82. Sketch of Mount Larrabee from The Pleiades, showing faulted slices of gneiss and schist of the Yellow Aster Complex in highly deformed shales of Chilliwack River terrane.

viewed from the point where the trail crosses a saddle on the south ridge of Mount Larrabee and skirts a steep rock buttress on the west (left); look north to The Pleiades to see blocks of light-colored gneiss of the Yellow Aster Complex on top of dark bedded sedimentary rocks.

90

YELLOW ASTER MEADOWS

Home of the Yellow Aster Complex

In the Yellow Aster Meadows area, west of Yellow Aster Butte, you can hike on a great slab of old gneiss of the Yellow Aster Complex (Plate 3A; Geologic Note 54). This fault-bounded block, some 4 square miles in area, is one of many in the Bell Pass Mélange (Chapter 3; Geologic Notes 54, 58, 59), though some blocks are only a few feet across. Where the rough and steep trail reaches the open, undulating country of the meadows and threads around bumps of gneiss and small ponds, look west to a rounded high knob of stacked thrust plates (Figure 83). Above the highly smashed gneiss holding up the meadows and ponds is a faulted layer of ocean deposits—chert and shale about 225 million years old (Triassic) that are also part of the Bell Pass Mélange. Above the chert and shale

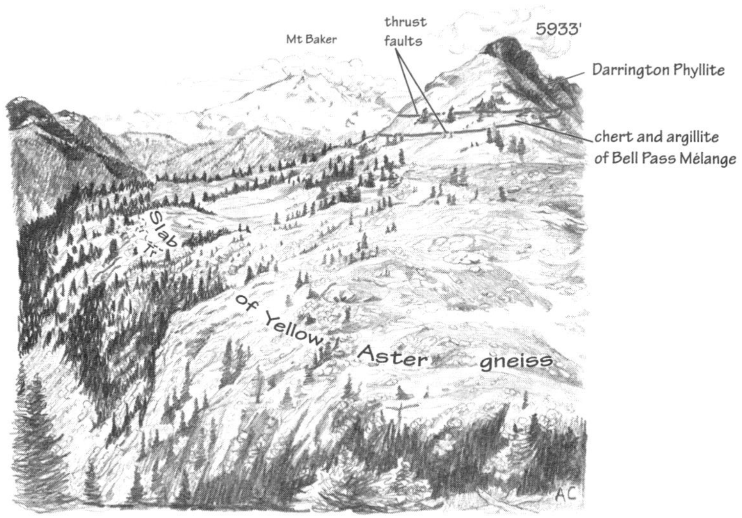

Figure 83. View of stacked thrust plates on thick, faulted slab of Yellow Aster gneiss holding up Yellow Aster Meadows (looking southwest from near Yellow Aster Butte).

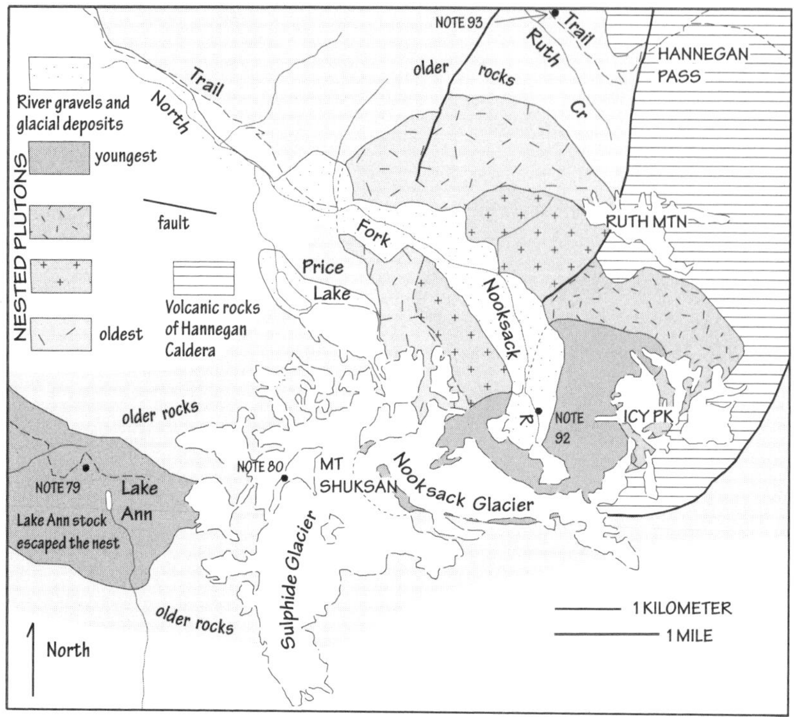

Figure 84. Young nested plutons of the Chilliwack batholith in Nooksack Cirque.

is Darrington Phyllite, another ocean-bottom deposit, about 160 million years old (Jurassic). Darrington Phyllite is part of the Easton terrane (Figure 15).

Miners' diggings here and there in the Meadows generally delve into pods of dark green serpentine, metamorphosed blocks of ultramafic rock from the Earth's mantle that have been dragged into the Yellow Aster gneiss along faults. The prospectors probably hoped to find nickel deposits, which are not uncommon in ultramafic rocks. They may have found some nickel ore, but unfortunately the miners knew nothing of mélanges, and did not realize that the small blocks they were digging in did not contain enough ore to make them rich.

91

MOUNT TOMYHOI

A tiny remnant of an overthrust plate

Climbers on Mount Tomyhoi will find that part of the broad plateau beneath the summit is completely different rock than that on the lower slopes of the mountain. A small patch of crinkled and smashed phyllite rests on top of bedded volcanic sandstones of the Chilliwack River terrane, the last erosional remnant of the upper thrust plate of the Easton terrane (Chapter 3; Plate 2 and Figure 15).

RUTH CREEK ROAD (USFS 32) AND BEYOND

92

NOOKSACK CIRQUE

Amphitheater carved from nested plutons

Hikers who reach the upper end of the North Fork of the Nooksack River and stand in the awesome amphitheater of Nooksack Cirque will experience an Ice Age scene of great grandeur. Perhaps, during the difficult hike over brushy, bouldery till and outwash that is necessary to reach the inner sanctum of the cirque, the hikers will have noticed a variety of granitic rocks in the outwash debris. The cirque, carved by numerous coalesced glaciers from east, west, and south, has breached a nest of igneous plutons of the Chilliwack batholith, mostly members of the Cascade Pass arc-root family (Figure 84). These nested plutons account for the variety of granitic rocks encountered in the area.

Geologists speculate that as a glob of molten rock intrudes upward from its origin, deep in the Earth above a subduction zone (Chapter 6), its outer margins cool, crystallize, and become rigid more quickly than the interior because they are in contact with the cool surrounding rocks

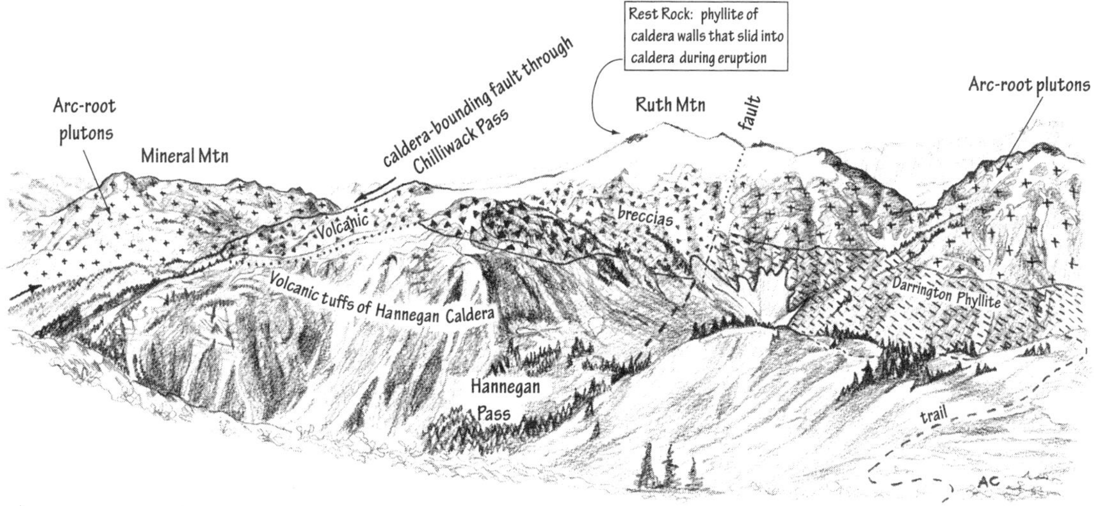

Figure 85. Hannegan Caldera as viewed from Hannegan Peak.

that the pluton is invading. While the margins solidify, the still molten core of the pluton continues to rise (Figure 31). Successive surges of the core, with accompanying cooling and crystallizing of the margins, form gigantic concentric tubes of granitic rock. If the molten rock feeding the system changes composition, then each thick-walled tube will be different from its neighbor and readily recognized as such. Moreover, the age of the tubes should decrease from the rim to the core. The plutons so boldly exposed in the Nooksack Cirque display this pattern, albeit imperfectly. We would expect the age differences from outermost to innermost tubes to be relatively small (on the order of 3 or 4 million years). In the case of the Nooksack nest, we know only that the pluton underlying Icy Peak is less than 4 million years old. Outside the nest, the nearby Lake Ann stock, which might be an offshoot of the youngest core pluton, is 2.2 million years old (Pliocene).

93

HANNEGAN PASS TRAIL

A view of Mount Sefrit, primordial magma of the Cascade Arc

As the trail to Hannegan Pass ascends the avalanche-scoured slopes of upper Ruth Creek, the dark cliffs of Mount Sefrit rise impressively across the creek to the southwest. Most of this mountain is underlain by a dark gabbro pluton of the Chilliwack batholith, a rock richer in iron, magnesium, and calcium than most of the other batholithic rocks. The Mount Sefrit gabbro is the slowly cooled, coarse-grained equivalent of basalt and could represent the primordial subduction zone melt that fed the Chilliwack batholith, but without the compositional changes induced by magmatic differentiation (Chapter 6). If the magma of Mount Sefrit had reached the surface of the Earth, it would have erupted as lava flows about 23 million years ago (Miocene). No basalt flows of that age are preserved in this area.

94

HANNEGAN PASS

Finding the age of an old volcanic caldera

The defile of Hannegan Pass is not very volcanolike, but the pass, in fact, has been eroded from the volcanic filling of ancient Hannegan Caldera. The eroded slopes all around are deposits of volcanic ash and breccia that erupted some 4 million years ago and filled a hole in the land created by the collapse of the roof of an emptying magma chamber (Geologic Note 77). The best view of the caldera and surrounding peaks is from the summit of Hannegan Peak, on the north. On the gentle, rounded top of Hannegan, the white chips and plates of volcanic rhyolite clink underfoot like pieces of pottery. The rock was originally mostly volcanic glass, chilled lava that was blasted into the sky and then fell back to Earth, still somewhat hot and tacky. As the pieces piled up, they flattened under their own weight. Now the glass is largely altered to clay minerals.

On the south, glacier-mantled Ruth Mountain is made of volcanic breccia (Figure 85). A black outcrop of rock sticking through the glacier near the summit of

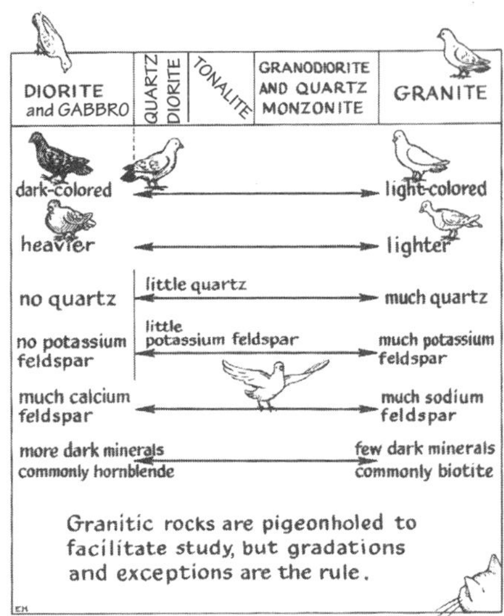

Figure 86.

Ruth—known to climbers as Rest Rock—is made up of black phyllite rubble, now cemented together. This mass of older rock is a remnant of a landslide that slid off the caldera walls when the volcano was erupting.

We know that this caldera erupted about 4 million years ago (Miocene) because Professor Joseph Vance has studied zircon crystals from the volcanic filling and determined their age by the fission track method.

The fission track method of dating a rock depends on radioactive decay, just as other radiometric methods do (Geologic Notes 4, 69), but goes at the problem in a different way. When radioactive isotopes of uranium atoms in zircon fission or decay, they emit atomic particles. As these particles pass through the crystal, they leave a tunnel of broken atomic bonds and other weaknesses in the structure that holds the atoms of the crystal together. The longer a crystal has existed, the more of these damage tunnels it will contain. A geologist who knows the amount of uranium, its rate of decay, and the number of damage tunnels can calculate the age of the crystal through a complicated process that includes cutting the crystals, etching them with acid to make the damage tunnels visible, and counting the number of such tunnels under a high-powered microscope.

95

COPPER RIDGE TRAIL

Hiking through gabbro

Good outcrops of black gabbro, rich in the mineral hornblende, can be seen where the Copper Ridge Trail zigzags from the lookout down to Copper Lake. This rock is older than the light-colored granite traversed south of the lookout and the light-colored tonalite and granodiorite exposed north of Copper Lake. Although all of these granitic rocks, including the gabbro, contain similar minerals, the amounts vary, and the proportions of the various minerals provide the basis for the classification system by which the rocks are named (Figures 86, 112).

Geologists use this classification to decipher the history of batholiths. For instance, study of the contacts between different igneous rocks of a composite batholith commonly shows that the denser, darker-colored gabbro formed first, followed by the less-dense, lighter-colored rocks. At Copper Lake, the gabbro and diorite, related dark rocks, are 34 million years old (Oligocene) making them components of one of the oldest plutons of the Chilliwack batholith, but similar rock holding up Mount Sefrit is 11 million years younger (Geologic Note 93). Both dark magmas represent little-differentiated subduction-zone melt (Chapter 6) that squirted up to the shallower crust of the Earth at different times.

96

MINERAL MOUNTAIN

Expanding granite and joints

Hikers descending the east side of Mineral Mountain will be impressed by the series of benches, like giant stairs, carved from the pinkish granite. Joints—that is, cracks in the granite—control the stairs in the landscape here. When the heavy burden of rocks over a deeply formed igneous pluton, such as this one, is removed by erosion, the unburdened granite expands and cracks, commonly forming a somewhat regular rectilinear pattern. Glaciers are particularly effective in quarrying large blocks bounded by joints, thus creating the steps. (To reach Mineral Mountain, see Tabor and Crowder, 1968, and Beckey, 1995.)

North Side Approaches

SKAGIT RIVER DRAINAGE

97

SILVERHOPE CREEK

An almost-nonexistent drainage divide

On the long trip up Silverhope Creek (not on the Geologic Points of Interest map, Plate 7A–D) and down the Klesilkwa River to reach the Skagit River and eventually Hozomeen Camp, travelers cross a drainage divide about 17 miles from the Trans-Canada Highway. Most travelers will not notice the divide in the gentle broad valley of forest and marsh. The Cordilleran Ice Sheet smoothed this broad valley, but evidence to the south (Geologic Note 5) indicates that once most of the rivers and streams south of this low pass drained through the Silverhope-Klesilkwa-upper Skagit valleys to the Fraser River. The drainage was reversed by the encroachment of the glacier, which filled the canyon and, on melting, drained south down the lower Skagit River.

HOZOMEEN CAMP ROAD AND BEYOND

98

HOZOMEEN CAMP

Tonalite and greenstone

Hozomeen Camp spreads itself on glacial deposits along the shore of Ross Lake. These gravels and sands overlie granitic rocks of the Chilliwack batholith. To examine the crystallized magma, follow the lakeside road south to outcrops of tonalite (Geologic Note 95 and accompanying Figure 86). The dark spots are biotite and hornblende. The light-colored crystals are quartz (glassy) and feldspar (white). This mass of molten rock, an arc-root pluton of the Snoqualmie family (Chapter 6), intruded the ancient volcanic rocks of the Hozomeen terrane about 24 million years ago (Miocene). A hike to Lake Hozomeen mostly traverses glacial deposits, but here and there in or near the trail are outcrops of dark green volcanic rock of the Hozomeen terrane. Much of this rock recrystallized as it was cooked by the heat of the intruding batholith (Geologic Note 102).

99

WILLOW LAKE

Drainage diversion

Just before reaching Willow Lake, the trail crosses a low drainage divide in a deep gorge carved from hard rocks of the Hozomeen terrane. Hozomeen Mountain rises abruptly to the north and Desolation Peak to the south. The feeble streams in this defile could not have carved such a passage through the hard rocks. This pass is a remnant of an earlier valley that drained westward from the upper parts of Freezeout Creek, lying to the east. The drainage was blocked by the Cordilleran ice and a series of stream captures have stolen all the waters, which now drain into lower Lightning Creek (Geologic Note 37).

ROSS LAKE BY BOAT

ROSS LAKE

Mariners' rocks

Ross Lake is an artificial lake held fast by Ross Dam (Geologic Note 12), but the valley it lies in was carved by glacial ice. Mountain views from a boat on Ross Lake are spectacular, but there are also things to see in the near view along the rocky shores. Many points of geologic interest accessible by trail from lakeside stops are described in sections on the trails (Geologic Notes 27, 28, 29, 30, 31, 35, 36, 37, 38).

SILVER CREEK

Building the alluvial fan

Alluvial fans form where a stream with the steep gradient of a side canyon meets the gentler gradient of the main canyon. Where the gradient of the side stream flattens, the stream slows down and can no longer carry the same amount of sediment. The dropped sediment builds up in a fan. The growing pile of sediment itself prolongs the steeper course of the side stream, allowing it to carry sediment farther out into the main valley. During times of high water, a stream on the fan may overflow its banks, especially at the head of the fan where the gradient changes and the stream is still dropping some of its sediment load. The new flood course will be a little steeper than the old course, and so the stream will reestablish in the new direction on the fan until that area too is built up like the old. Over a period of time, the stream swings back and forth across the fan.

The Silver Creek fan is the most prominent fan along Ross Lake today. Farther south along the lake, other fans have been covered by the lake, and sediments now accumulating at the bottom of the lake, dumped by the side streams, have yet to build up above lake level (Geologic Note 117).

LITTLE JACKASS MOUNTAIN

A geologic contact where magma meets its container

From the lake, and with good western sunlight, a hard look at the west side of Little Jackass Mountain reveals light-colored granitic rock in the lower cliffs, and dark volcanic rock in the upper cliffs. The contact between these contrasting rocks can be followed—by eye anyway—across the precipitous slope. The granitic rock of the Chilliwack batholith (Geologic Note 98) intruded the volcanic rock above. This view exhibits in cross section part of what was once a vast underground reservoir of melted rock (a magma chamber) and its roof (Chapter 6).

WEST SHORE OF ROSS LAKE

Radiolarian chert

Along much of the western shore of Ross Lake, especially from opposite Jack Point to Arctic Creek, lakeside outcrops are radiolarian ribbon cherts of the Hozomeen terrane. Look for the tell-tale bedding characterized by one-to-three-inch layers of dark, hard, somewhat glassy or stonewarelike quartzite, separated by flaky gray, and much softer, shaley layers (Figure 87). (For more information on radiolarian chert see Geologic Note 30.)

CLIFFS OF NO NAME CREEK

Slumps and slides under the Hozomeen ocean

Cliffs at the mouth of No Name Creek reveal catastrophic events on the bottom of an ancient ocean. Fragments of chert, basalt (now greenstone), limestone, and other rocks of the Hozomeen terrane are surrounded by shale. Such mixing of rock types can form in many settings, but here geologists think that the mixture formed by submarine landsliding down the flanks of a 190-million-year-old (Triassic) oceanic volcano. Similar but much larger underwater slides have been found on the flanks of modern oceanic volcanoes in Hawaii.

DEVILS CREEK GORGE

Hanging valley defile

A boat excursion into the mouth of Devils Creek is particularly enticing. The lake has flooded a winding narrow defile, an erosional slot carved by Devils Creek

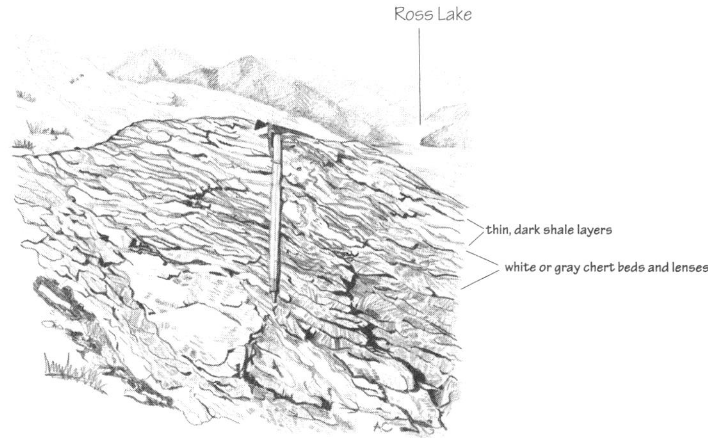

Figure 87. Banded radiolarian chert on ridge above Skymo Creek looking north to Ross Lake. During deformation, the light-colored chert beds have been broken and stretched between thin, plastic, shale layers.

into volcanic rocks of the Hozomeen terrane. Many tributaries to the Skagit have hanging valleys, high-level glacial troughs carved when their small alpine glaciers could not cut down as fast as the major trunk glacier in the Skagit valley (Geologic Note 31). The smaller side streams of these alpine valleys still leap out in waterfalls to descend the glacier-carved step at the valley side, but powerful Devils Creek has sawed rapidly down, creating the lake-level slot.

106

ROLAND POINT

Where magma could not digest old mantle

The south side of Roland Point offers a look at rock pudding, in which outcrops of light-colored granitic rock are rich in rounded and angular fragments of dark rocks. This igneous breccia represents a place where invading magma surrounded older fractured rocks, which were engulfed as fragments in the molten rock. Some of the captured pieces are aggregates of the mineral hornblende, and some are ultramafic rocks. Iron and magnesium-rich ultramafic rocks melt at very high temperatures only. Since granitic melts are not hot enough to melt these dark rocks, pieces of mantle caught up in granitic plutons survive their hot bath. (One of the authors would like to note that Roland Point appears to have been named for Tommy Rowland, who homesteaded on the Skagit River in the late 1800s. Tommy Rowland spelled his name in the proper way. Deviant cartographers have altered the spelling.)

South Side Approaches

SUIATTLE RIVER DRAINAGE

107

SUIATTLE AND SAUK RIVERS

Glacial silt

Just south of the bridge carrying State Route 530 across the Sauk River, look at the color change in the river water. During high runoff, the mud-choked Suiattle River can be seen hugging the eastern bank of the clear green Sauk. Most of this mud is washed from silt, sand, and gravel deposits of the Great Fill (Geologic Note 114) on the east side of Glacier Peak volcano. Even when the water is low and the contrast less marked, the Suiattle brings in glacial silt that colors the Sauk for miles below the junction.

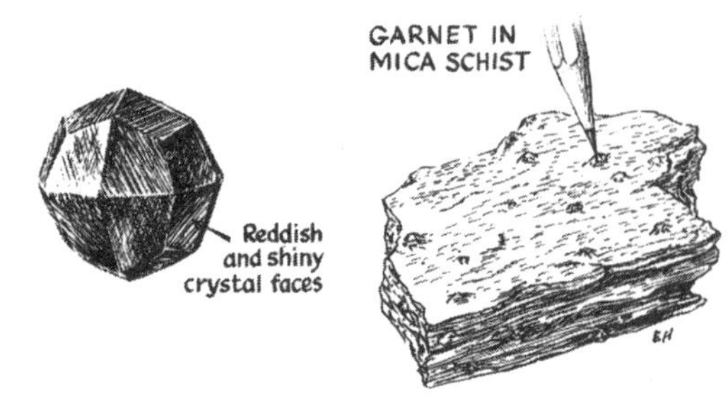

Figure 88.

108

GREEN MOUNTAIN TRAIL

Signs of life in schists of the Nason terrane

About halfway up the Green Mountain Trail, hikers encounter outcrops of metamorphosed shales and tuffaceous shales, now garnet-mica schist and hornblende-mica schist of the Nason terrane. These rocks began in the ocean, probably at the toe of a submarine fan built of debris eroded from a volcanic arc (Chapter 3; Figure 19). A close look at the rocks may reveal tiny red garnets (Figure 88). Much of the schist is dark with graphite, a metamorphic form of carbon, indicating that the original shale was rich in organic debris. Below the small lake, graphite colors the rocks gun-metal gray.

109

SUMMIT OF GREEN MOUNTAIN

Handsome sill and rusty red mountain

At the summit of Green Mountain (4 miles and 3,000-foot elevation gain) are 360° views of the Glacier Peak Wilderness that are seldom seen so close to a road (Figure 89). The summit is supported by a handsome sill of greenish andesite (Geologic Note 7), a young igneous intrusion into the enclosing schists. The same sill is visible on the grassy face of the next peak to the north. Look closely at the rock to see black squarish crystals of hornblende and white rectangular crystals of feldspar, about the size of lentils, scattered in a finer-grained crystalline matrix. The larger crystals formed early in the magma chamber that fed this sill.

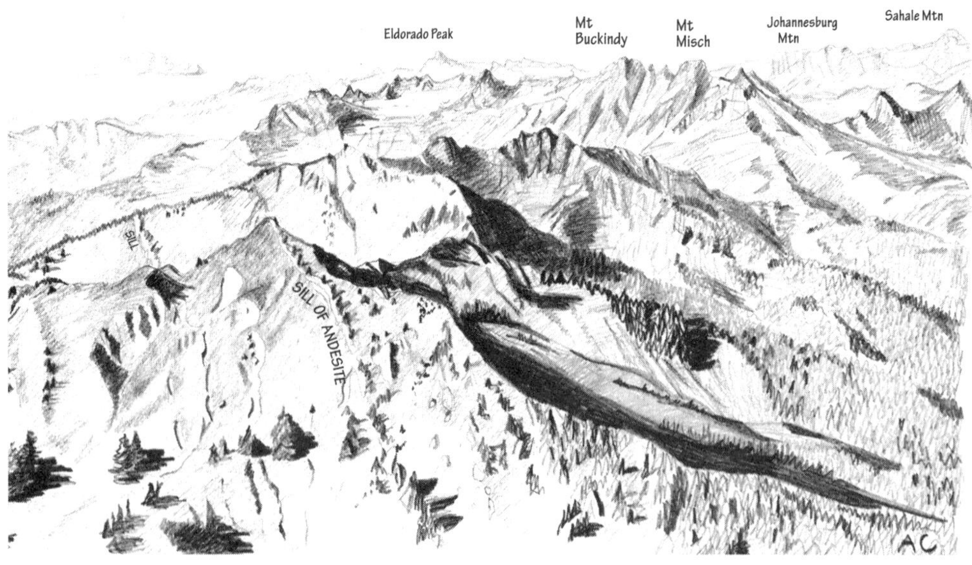

Figure 89. View to the north from Green Mountain. Ribs held up by resistant sills of andesite contrast with the meadowy slopes eroded from schist. Layers and foliation in the schist and the sills are standing almost vertically.

Figure 90. Talus.

Look north-northeast to the rusty red spires of Mount Misch, where the gray schists and gneisses of the Metamorphic Core Domain have been fractured and basted with the juices emanating from a nearby Cascade Arc pluton. The resulting rock is rich in pyrite, an iron-sulfur mineral, which weathers to produce sulfuric acid. The acid attacks the rock to produce clay minerals that are stained orange-brown by iron.

110

SULPHUR HOT SPRINGS

The smell of a volcano

A short hike up Sulphur Creek (on a desultory trail from the Suiattle River road) may afford a look at the disappointingly small seeping of warm sulfurous water known as Sulphur Hot Springs. Odoriferous gas (hydrogen sulfide) may lead hikers to the site. Sulphur Hot Springs is the smallest of the known hot springs near Glacier Peak and the most distant from it. The others are Kennedy Hot Springs and Gamma Hot Spring (both are south of the area covered by this guide).

111

SULPHUR MOUNTAIN

Crystals of pyroxene in the Sulphur Mountain pluton

Hikers in the alpine country of Sulphur Mountain or Bath Lakes will see many coarse talus slopes, mostly clean blocks of granitic orthogneiss that have fallen from small cliffs (Figure 90). A large mass of granodiorite, the Sulphur Mountain stitching pluton (Chapter 5; see Plate 2), makes up the ridge here, and the many joints in the rock contribute to the blocky debris. Water freezing in the joints separates the blocks. The expanding ice slowly deconstructs the mountain.

A close look at the rock reveals fingernail-sized spots of glassy quartz and uncommonly large stubby pencils of light-green pyroxene surrounded by white feldspar (Figure 91). Pyroxene is a complex mineral common in igneous and some metamorphic rocks. This rock was once a molten mass, which was squeezed and recrystallized during metamorphism. In the center of the pluton the squeezing is not so obvious, but the eastern margin of the pluton, many miles east of Sulphur Mountain, is a strongly foliated rock (Figure 55).

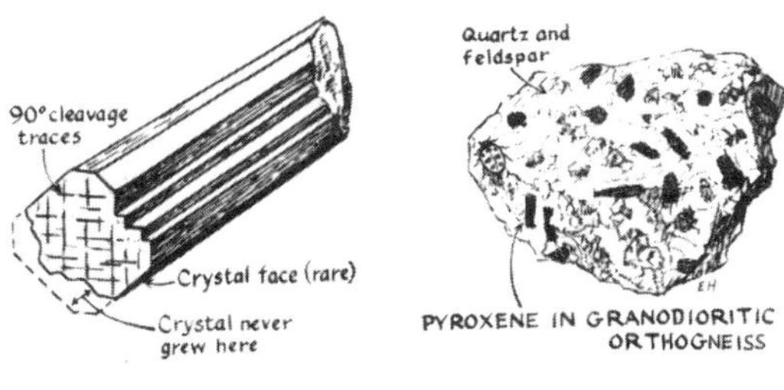

Figure 91. Pyroxene crystal.

SUIATTLE RIVER AND SUIATTLE PASS TRAILS AND BEYOND

112

SUIATTLE RIVER

Glacial terraces

For several miles, the Suiattle River Trail follows narrow terraces (Figure 92) made of river gravels. The Suiattle River at one time could not carry all the debris delivered to it by the side streams and glaciers, so it deposited the surplus in the valley (Chapter 7). Now, when the load of debris is less, the river has cut down through the gravel into underlying rock and carried most of the terraces away, leaving only small patches of gravel here and there. At one place (1.8 miles from the trailhead), the trail crosses a "slide" (a steep river-cut in one of the terraces) composed almost entirely of volcanic material washed down from Glacier Peak.

113

JUST BEYOND CANYON CREEK

A dry hillside eroded from granitic rock

Upvalley from Canyon Creek (5.7 miles and a 700-foot elevation gain from the trailhead), small pines and ferns grow on the dry hillside in contrast to the lush forest downriver. The hillside is out of the frequently shadowed gloom north of Grassy Point and in the rain shadow of Glacier Peak. But the dryness is also caused by the nature of the rock. Just below Canyon Creek, the trail leaves schists and gneisses and enters granodiorite. The granodiorite is jointed and weathers to a coarse sand, or grus (Geologic Note 45). Water seeps away rapidly through the joints and sand, leaving the soil dry. Hikers can enjoy views of granodiorite walls at openings where the trail leaves the forest to cross rocky draws, which are also quite dry by midsummer.

The rocky draws also provide views southwest across the river, where a rocky face surmounts forested slopes. The blocky jointed cliff is a truncated lava flow from Glacier Peak volcano; the flow caps a spur made of granodiorite (Figure 86). The granodiorite is part of the Cloudy Pass batholith, a pluton of the Cascade Volcanic Arc (Chapter 6), which invaded the gneiss and schist of this area about 22 million years ago (Miocene).

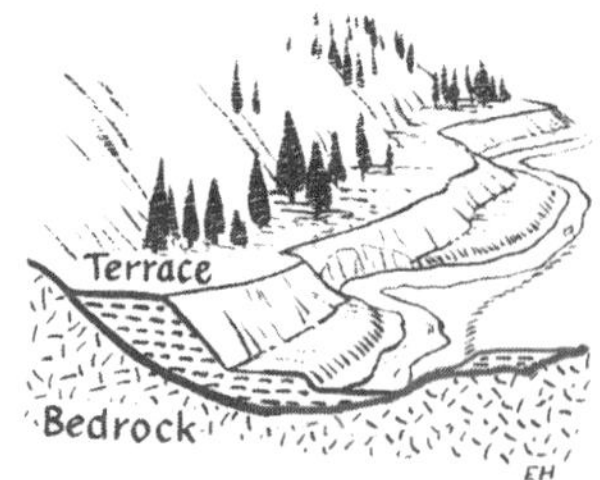

Figure 92. River terraces.

MINERS CREEK

The Great Fill

In the vicinity of Miners Creek (9.7 miles, 1,000 feet of climbing), hikers on the Suiattle River Trail can look west across the river to a bluff of light-colored silt, sand, and gravel. The bluff is a remnant of the Great Fill[1], once a thick, valley-filling mass of debris from Glacier Peak. According to the volcanologists, many, many floods of volcanic debris came down from Glacier Peak to fill the Suiattle canyon between 6,700 and 3,400 years ago. Most is eroded except for that making up a huge ramp on the west side of the peak and a wide terrace in the upper Suiattle valley (Figure 93).

[1] Recipe for Chicken in the Great Fill (courtesy of 1977 Darrington Ranger District trail crew). Dig a one- to two-foot-deep hole in dry sand of the Great Fill. Build a fire in it and allow to burn for one hour. While waiting, rub the inside of a whole roasting chicken with oregano, basil, salt, pepper. Stuff with sage-rich bread dressing or onions and mushrooms. Place stuffed chicken in a three- to four-quart cook-pot and surround with vegetables—onions, garlic, mushrooms, carrots, quartered potatoes. Cover with larger inverted pot. Place shovel of sand on fire. Place pot with chicken on sand. Collapse walls of fire pit, smothering fire and covering pot. Build new fire on top and allow to burn for one hour. Exhume and eat.

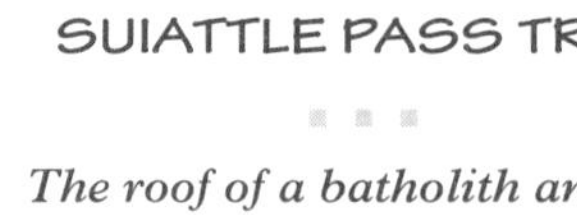

SUIATTLE PASS TRAIL

The roof of a batholith and the Glacier Peak Mines

Hikers on the Suiattle Pass Trail may be surprised on the traverse of the hillside above Miners Creek where, near the junction with the Pacific Crest Trail, they come across the remnants of a mining camp (6.2 miles from the

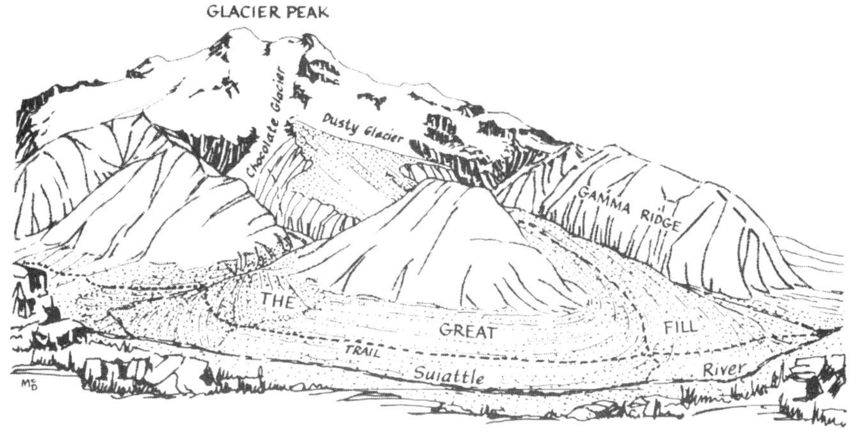

Figure 93. The Great Fill as seen from the east.

Suiattle River Trail, 2,600-foot climb). For many years the Bear Creek Mining Company, a subsidiary of Kennecott Corporation, based an intensive prospecting operation on Miners Ridge. Company geologists have explored an extensive but low-grade copper deposit and drilled many holes into the mountainside to sample the deposit. Because the claims were made and patented before the establishment of the Glacier Peak Wilderness, the company has had access rights. The encampment here, once quite extensive, was supplied by helicopter for many years. For a heated discussion of wilderness values and mining at this site, see McPhee, 1971.

Not so noticeable here, but obvious when viewed from the south, these slopes below Suiattle Pass are big ramplike benches (Figure 94). These are controlled by widely spaced joints in the granodiorite, which underlies the pass and lower slopes of Miners Ridge. Plummer Mountain and the crest of Miners Ridge out to Image Lake are carved in shattered brown gneiss lying over the granodiorite. The top of the once-rising mass of melted rock stopped here. The scratchings of the prospectors on Miners Ridge can be seen along this roof of the Cloudy Pass batholith (see Plate 2; Figure 30), where fluids from the once-molten rock concentrated the ore minerals (Geologic Note 34).

Geologists once thought that the brown gneiss capping the granodiorite was one of the oldest rocks exposed in the North Cascades. It is biotite gneiss of the Swakane

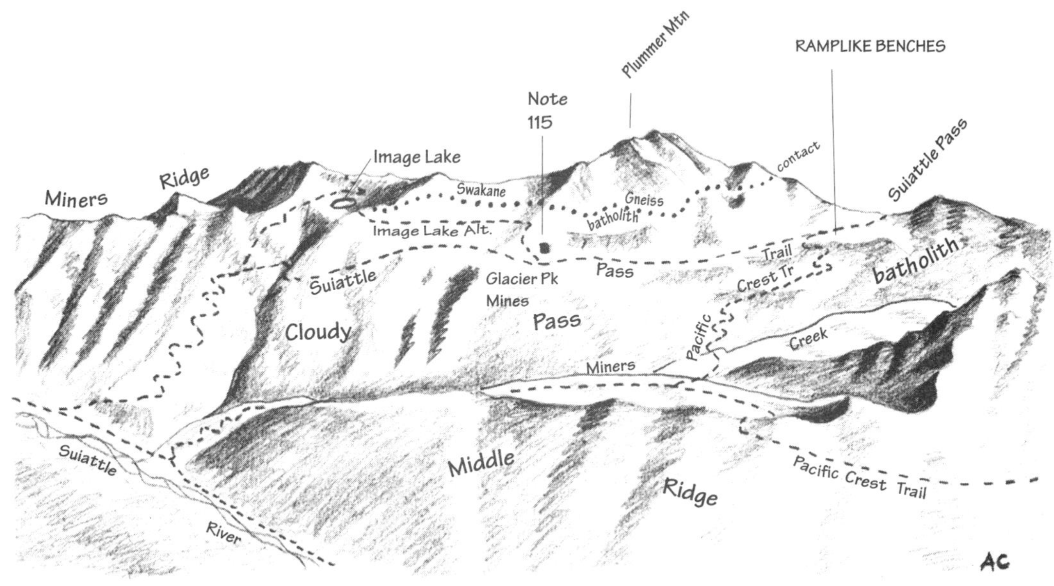

Figure 94. Looking north at Miners Ridge. Swakane Gneiss roofs the Cloudy Pass batholith along the contact (dotted line).

terrane, which has had an unusual geologic history, and is now thought to be one of the youngest terranes in the Metamorphic Core Domain (see Chapter 3).

116

CANYON LAKE TRAIL

The tites, migma and pegma

En route to Canyon Lake, lucky hikers can enjoy smooth expanses of glacier-polished biotite gneiss of the Swakane terrane (about 3 miles from the junction with the Image Lake Trail). The smooth rocks are complex mixtures of brownish and streaky white biotite gneiss that have been cut and sliced by younger white dikes. The forms of the dikes in this migmatite (Geologic Notes 11, 20) are fantastically varied—crooked and straight, bent and swirled, short and fat, long and streaky. The streaks in the gneiss are also bent and swirled. The whole mass looks like a pastry much stirred by some fantastic spoon. Surely, the whole mass was never molten, or the different rocks would have melted together.

A search will show that some of the white dikes have particularly large crystals of white, milky feldspar. These have shiny smooth breaks (cleavage) which reflect the light. Also common is the black or brown mica, biotite, in stacks or thin flakes; less easily found is rare silvery white muscovite. Dike rocks with large crystals like these are called *pegmatites,* and in many regions, but not here, they contain some of the rarest of minerals, some of which are occasionally of gem quality.

SOUTH SIDE APPROACHES

LAKE CHELAN AND STEHEKIN RIVER DRAINAGE

LAKE CHELAN, THE STEHEKIN RIVER ROAD, AND BEYOND

117

LAKE CHELAN VALLEY

Remarkable glacial erosion

The rounded bluffs and huge grooves on rock slopes of the lower Lake Chelan area are ample evidence that glaciers sculpted the valley. The lake elevation is about 1,100 feet above sea level. The ice cut deeply. In one place, about 29 miles up-lake from Chelan, the bedrock bottom is 436 feet below sea level. On a hot day, boaters may find it difficult to visualize the 5,000-foot thickness of ice that once filled the canyon and engulfed many of the surrounding peaks and ridges. The only toeholds man has established on the steep shores are on gravelly

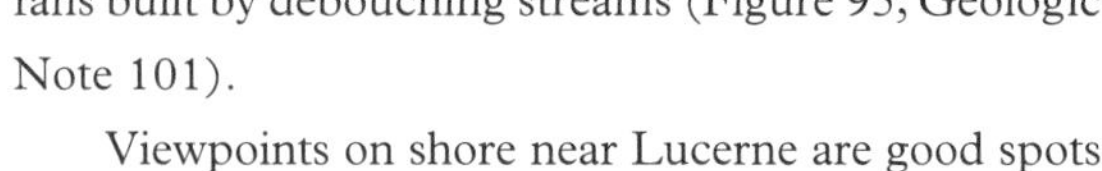

fans built by debouching streams (Figure 95; Geologic Note 101).

Viewpoints on shore near Lucerne are good spots to conjure images of the glacier that carved Lake Chelan and Domke Lake. The summits of both Domke Mountain and Round Mountain, across the lake, are rounded and smoothed by the ice, indicating that glacial ice in Railroad Creek, Fish Creek, and Lake Chelan filled the valleys to elevations above these summits (Figure 96).

Figure 95. Alluvial fan on Lake Chelan at Railroad Creek.

118

LAKESHORE TRAIL AT HUNTS BLUFF

Traces of the glacier

Glacially scoured bedrock at Hunts Bluff is a reminder of the mile-thick ice that

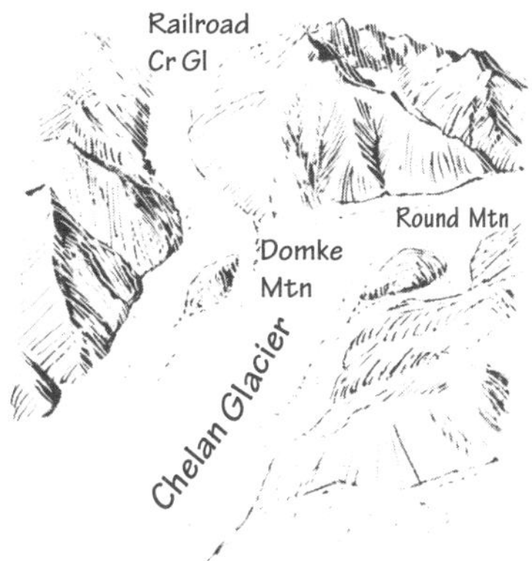

Figure 96. At the point of maximum ice cover, Domke and Round Mountains were completely overidden by the glacier.

filled the Chelan valley during the last ice age, about 20,000 years ago. The rounded, irregular landscape is typical of ice scour. At the north end of the bluff, on the west side of the Lakeshore Trail, sparse glacial striae cut across the ice-polished orthogneiss. These sets of straight, parallel scratch marks are typically only a fraction of an inch across and up to several feet long. Weathering of the rock has destroyed most of the polished, striated surface in this area.

A deposit of sand and gravel overlies the bedrock in places. Unlike the usual disorganized, clay-rich mess left by melting ice, the sand and gravel in this deposit are sorted into distinct beds. The sand and gravel were sorted by flowing water, probably by a stream that flowed along the margins of the glacier in the Lake Chelan valley as the ice melted and thinned at the end of the last ice age. The contact between the deposit and the underlying gneiss is what is known as an unconformity (Geologic Note 146), although it has little chance of being buried, changed to hard rocks, and becoming part of the rock record for examination by geologists of the future.

119

PAINTED ROCK AT STEHEKIN

Rock art

Painted Rock, across from the Stehekin boat landing, bears Indian paintings of unknown age. These petroglyphs were described for the first time in the late 1800s (Figure 97). At that time the figures of men and animals were about 17 feet above lake level. Since then, Lake Chelan has been raised artificially about 21 feet to facilitate power generation at Chelan, and some of the paintings are now under water. How the Indians painted so high on the steep rock is unknown, but possibly the lake level had changed before, having been higher perhaps at some time before the arrival of Europeans (Geologic Note 120).

120

RAINBOW FALLS

Accidental birth of the falls

Much-visited Rainbow Falls is an anomaly in the Stehekin valley. Out of many, many side tributaries to the Stehekin

Figure 97. Petroglyphs near Lake Chelan (drawn from a photograph).

After glaciation, Rainbow and Boulder Creeks have notched old, relict valley-bottom terrace.
Alluvial fans develop and cover lower cliffs of old valley-bottom terrace.
Rainbow Creek
Boulder
Creek
relict of old fan
Old valley bottom terrace
FALLS
trail
Stehekin
road
River
AC
To Stehekin
Fans eroded away by meandering Stehekin River. Rainbow Creek gets caught by bedrock lip.

Figure 98. Development of Rainbow Falls.

River, Rainbow Creek is the only one with a significant waterfall that plunges right into the flat-floored main valley. For example, Rainbow Creek and Boulder Creek both have upper hanging valleys, left high by the more vigorously downcutting Stehekin valley glacier (Geologic Notes 31, 105). Both also have much the same discharge of water and are eroding the same hard rock of the Skagit Gneiss Complex. Rainbow Creek, however, plunges 240 feet over a sheer cliff, while Boulder Creek cascades down a gorge eroded back into the lip of the hanging valley. Moreover, upper Rainbow Creek is also in a gorge, above the junction of the Rainbow Lake Trail with the Rainbow Loop Trail. A probable explanation for the difference between these two streams is found in the remains of an old alluvial fan of Rainbow Creek. The fan is perched on a remnant of an old valley bottom (Geologic Note 125), which is now a rock bench situated about 1,000 feet above the present Stehekin valley floor. Both the bench and the fan are readily appreciated from the Rainbow Loop Trail, which winds along the bench and climbs over remnants of the fan (Figure 98).

Between major glaciations, both creeks had indeed cut gorges, as the rushing water sawed into the lips of the hanging valleys. After the last major ice advance down the Stehekin valley, however, both creeks built large alluvial fans, which buried their lower gorges, as well as the lower valley cliffs below the remnants of the old valley bottom. These old fans temporarily built out on the main valley floor and pushed the Stehekin River to the western side of the valley. The lower part of Rainbow Creek's gorge was probably southeast of the present Rainbow Falls and more in line with the creek's trend above the falls. But a big river like the Stehekin never stops its attack on valley-bottom debris, and it gradually ate away at the old fans, eventually cleaning most or all of them away, except for the parts perched on the rock bench beyond its reach or buried by the present-day, newly forming fans. When the Stehekin River had cut back the toes of the old fans significantly, Rainbow and Boulder Creeks, falling over the steep cut, began to saw into their respective fans.

Creeks building alluvial fans tend to wander back and forth on the fans (Geologic Note 101). Boulder Creek found its old gorge as it cut down, but Rainbow Creek had wandered north and more or less by chance soon sawed down into a fresh rock lip at the edge of the bench, the site of the present Rainbow Falls. The creek cannot get out of its new channel cut in bedrock above the falls, at least until it builds its fan up to its former dimensions. The old channel remains buried under the old fan.

121

HIGH BRIDGE

Agnes Gorge and the upper Stehekin gorge

Above High Bridge (10.6 miles from Stehekin), the Stehekin River Road enters a rocky canyon cut into the lower end of a hanging valley of the upper Stehekin River. Agnes Creek, a major tributary branching off to the west, has carved a similar deep canyon. At this point the glacier flowing down the Stehekin valley was augmented by ice from Agnes Creek, and the increased eroding power left both the upper Stehekin and Agnes Creek hanging. The inner gorges here probably have been cut since the glaciers retreated about 14,000 years ago. The bench above the gorge is an old valley bottom, which was probably carved in an earlier glaciation (Geologic Notes 120, 125) Exposures along the road here are orthogneiss of the Skagit Gneiss Complex (Chapter 4).

TRAILS REACHED FROM THE STEHEKIN RIVER

122

PURPLE PASS

Orthogneisses

The climb to Purple Pass is long and grueling (8.4 miles; 5,800 feet), but the views of the glacier-carved, fjordlike canyon of Lake Chelan are splendid even part way up the trail. The rocks along the trail are streaky orthogneisses of the Skagit Gneiss Complex. Most hikers will not want to examine closely the varieties of orthogneiss displayed on this huge mountainside, but zealous geologists have, and they portray the mountain as a packet of stretched igneous plutons that were once deep in the Earth's crust (Chapter 4; Geologic Notes 7, 124). Just to imagine the amount of hard rock removed by erosion here is effort enough during the hard climb.

123

COMPANY CREEK TRAIL

Relict rounded rocks and a view of Bonanza Peak

Climbing up the first few miles of the Company Creek Trail, you may wonder about the rounded stream boulders and cobbles now perched along the trail, high above any present-day streams. The gravels shedding these stones may have been deposited by Company Creek alongside the glacier that filled the Stehekin valley about 15,000 years ago (late Pleistocene). But, since geologists cannot see the shape of the beds in the gravel deposit, only its debris, weathering out of the slope, they should consider an alternative theory: namely, that the boulders and cobbles could be remnants of a large alluvial fan that formed shortly after the glacier withdrew and were later mostly eroded away by the Stehekin River (Geologic Note 120). Company Creek itself has sawed a slot in the bedrock since the glaciers retreated. As with many tributaries to major valleys, the small side-stream glacier that smoothed out Company Creek could not keep up with the Stehekin valley glacier, leaving a hanging valley (Geologic Notes 31, 105) when the ice melted.

This upper level of the Company Creek valley can be most appreciated from a dramatic viewpoint where the trail leaves the stepped ridge and turns a corner onto the steep Company Creek valley wall (about 1.5 miles from and 1,700 feet above the trailhead on the west-side road). Look west to see the broad glacier-carved, forested upper valley of Company Creek. At the valley head, Bonanza Peak bears the Company Glacier on its north side (Figure 99). Bonanza, at 9,511 feet, is the highest nonvolcanic peak in the North Cascades. It is carved from hard, metamorphosed tonalite of a Marblemount pluton (Chapter 3; Geologic Note 68).

Beyond this viewpoint, the trail crosses outcrops of light-colored Skagit Gneiss Complex and eventually reaches a talus slope, where blocks exhibit varied orthogneisses and pegmatites of the complex. Many blocks are migmatite (Geologic Notes 11, 20, 116). Beyond, few geologic sights reveal themselves in the miles and miles of brushy trail.

124

McGREGOR MOUNTAIN TRAIL

A million switchbacks through lineated orthogneiss to a splendid view

Pause on the small moraine below the final rock scramble to the summit of McGregor Mountain (about 7.5 miles from and 5,700 feet above the trailhead at Highbridge) and consider the rocks displayed along the endless switchbacks of the trail below. The rocks belong to the Skagit Gneiss Complex (Chapter 4) and were born many miles deep in the Earth's crust. Most of the outcrops are of strongly lineated orthogneiss. (Geologic Notes 7, 122). Some of the lineated rock is light-colored, fine-grained granite, which, because of its small crystals, does not reveal its mineral alignment except through close examination. Some of the fine-grained rock is dark tonalite. A small amount of the rock is coarser-grained and foliated, indicating that it is the older of the solidified and recrystallized magmas in the complex (see "Orthogneiss and Migmatites" in Chapter 4).

Large, white-speckled black blocks in the moraine are neither foliated or lineated. They are from diorite dikes that invaded cracks in the Skagit Gneiss after it had been uplifted many miles.

Scramble up the precipitous, arrow-marked trail to the summit of McGregor to see more of the orthogneiss and, even better, to view myriad peaks in the Metamorphic Core Domain to the north and west, and many more in the Methow Domain to the east (see Plate 2). This summit provides one of the region's most impressive views.

Figure 99. Bonanza Peak (9,511 feet) at the head of Company Creek. On the left is the end of Sable Ridge.

125

GOODE RIDGE TRAIL

Views of valley evolution

Where the Goode Ridge Trail leaves the forested lower flanks of the ridge and switchbacks up steep meadows (about 2.5 miles and 2,300 feet of climbing from the trailhead), views of Bridge Creek reveal an old valley floor incised by a newly carved inner gorge (Figure 100). Much of the old valley floor is preserved downvalley as far as High Bridge on the Stehekin River. Below there, only remnants can be found along the valley sides such as at Coon Lake and near Rainbow Falls (Geologic Note 120). Geologists do not know exactly when this old valley floor was made, but it was probably carved by ice during some glaciation preceding the most recent one. During this last glacial advance, ice modified the old floor only slightly in valleys above Highbridge, while all but devouring it below Highbridge.

126

NORTH FORK BRIDGE CREEK TRAIL

Thoughts at the base of Goode Mountain

About a mile above Grizzly Creek, the North Fork Bridge Creek Trail emerges from endless avalanche brush into a small, boulder-strewn meadow directly across from the northeast face of Goode Mountain. This is a good place to pause and contemplate the many ways mountains begin their voyage to the sea.

The meadow is growing on an alluvial fan. The fan is being built by slurries of water, clay, sand, and boulders that the occasional cloudburst or rapid snowmelt washes down the gully at the head of the fan. The slurries spread out across the valley floor. Farther upvalley and to the north is a small talus—the word refers both to the angular bouldery debris and the slope that it makes—of migmatitic gneiss blocks that have fallen off

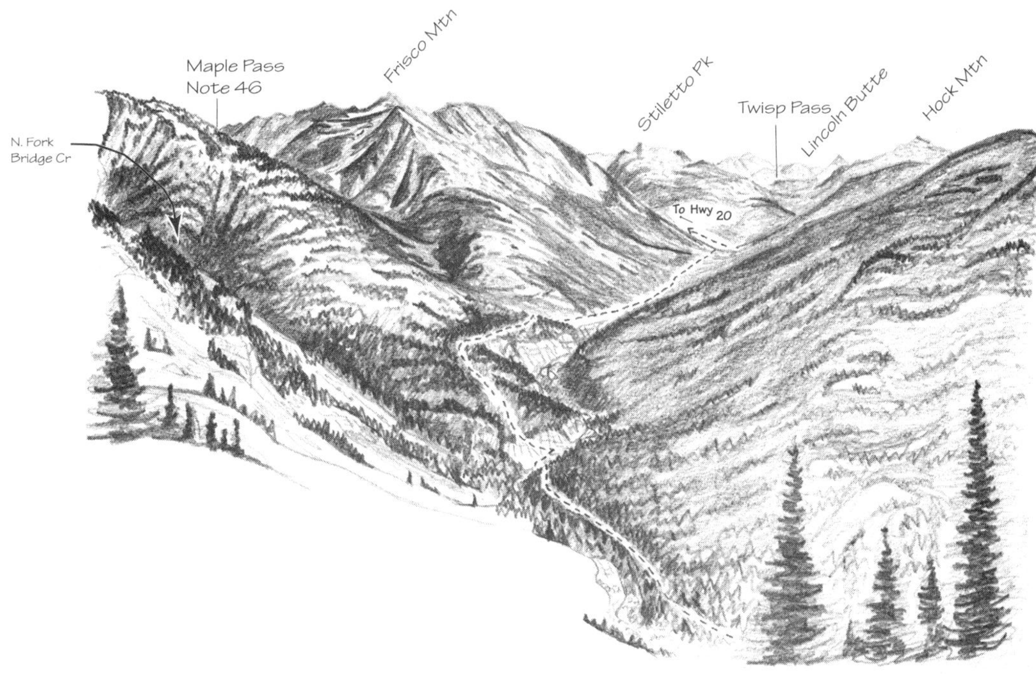

Figure 100. View eastward up Bridge Creek from the Goode Ridge Trail. Bridge Creek has cut a gorge in the bottom of the more gently floored valley.

the cliff above, either pried loose by water freezing in cracks, kicked loose by mountain goats, or shaken off by earthquakes. Some of the blocks have rolled out onto the fan, where they make good perches for geologizing visitors. Across the valley, halfway up Goode Mountain, a small balcony glacier is slowly enlarging its niche, plucking blocks from its bed and grinding away. Hikers who wait long enough may hear the roar of an ice block (and its contained rock) falling off the glacier and disintegrating on the slabs below. Blanketing all is the unceasing roar of the North Fork, tumbling boulders down its bed. These mountains won't last forever.

But how long will they last? Studies measuring how much mud is carried by rivers and how fast reservoirs silt up suggest that regionwide erosion lowers the landscape at rates of roughly one inch per century. Goode Mountain rises almost 6,000 feet above this spot, so destruction of this mountain will probably take millions of years.

127

CASCADE PASS TRAIL

Old mining junk

An overgrown meadow about 1.2 miles up the Cascade Pass Trail from Cottonwood Camp hides the rusted remains of a Horseshoe Basin mining enterprise, dating back to the late 1800s. A description in the *Chelan Leader* from that time, by De Witt Britt, is full of unbounded optimism for development of rich lead deposits in Horseshoe Basin. The mineral-laden solutions that left galena and other minerals in Horseshoe Basin, as well as near Cascade Pass, in Park Creek, and along the North Fork of Bridge Creek, probably emanated from the crystallizing melt of the large Cascade Pass dike. The magma of the dike filled a northeast trending crack, which is over the main ridge to the north. The region of mineralization parallels the dike (Geologic Notes 26, 64; Figure 30).

128

HORSESHOE BASIN TRAIL

Crystal alignment or going with the flow

On the old road to Horseshoe Basin, just south of the junction with the trail to Cascade Pass (2.1 miles, and 900 feet of climbing above Cottonwood Camp), a clean cliff of Eldorado Orthogneiss can be readily examined. This old stitching pluton invaded the metamorphic rocks of the Chelan terrane about 90 million years ago (Cretaceous). Note the prisms of black horn-

Figure 101. View looking south from near Doubtful Lake of Magic Mountain Gneiss thrust over Cascade River Schist of the Chelan Mountains terrane.

blende (Figure 40), which are commonly aligned in the rock. Although metamorphic squeezing can align flat minerals like micas into a platy structure (foliation), or elongated minerals into a linear ordering (lineation), geologists think that much of the hornblende in this rock was aligned by the flow of the molten rock when it moved into place prior to being metamorphosed (Geologic Note 7).

129

PELTON BASIN

View of the Magic Mountain Gneiss

Where the trail climbs onto talus in Pelton Basin (3.8 miles and 1,900 feet of climbing above Cottonwood Camp), pause for good views of the Magic Mountain Gneiss, on top of schists of the Chelan Mountains terrane (Figure 101; Geologic Notes 70, 72). Pelton Basin is the upper cirque (really a hanging valley) of the double-cirqued Stehekin River (Chapter 7).

RAILROAD CREEK TRAIL AND BEYOND

130

HOLDEN VILLAGE

The costs of mining

Charming Holden Village is composed of homes once built for the Holden Mine administrators. The village now houses staff and guests of the Lutheran Church, which operates a retreat center (open to all, but by reservation only). Holden Mine prospered for about 20 years as the miners drilled and blasted in the flank of Copper Peak (Figure 102). When the mine was in operation, great buckets of ore concentrate, containing copper, some zinc, and a little gold were trucked down to Lucerne and placed on barges. The concentrate was put on trains at Chelan and taken to Tacoma for smelting.

The mine closed and staff and miners left in 1957, not because the metals ran out, but because of excessive mining costs. The deepest mine workings are a half-mile below creek level. The labyrinth of passageways and the huge caverns where blocks of ore were removed are now filled with water. High on the mountainside, above the remains of buildings at the main entrance level, small yellow dumps mark entrances to the uppermost and still- dry galleries. The huge amount of rock removed and

Figure 102. Holden Mine and village about 1951 (viewed from the northeast).

pulverized is depressingly demonstrated by the tailing dump. Chemicals used to separate the ore minerals remain in the tailing and hinder much plant growth. For years the wind lifted clouds of this pulverized rock and dusted it over the valley, while Railroad Creek carried slime to Lake Chelan to color the boulders there a striking rusty orange. In 2012 work began to clean up the now-identified Superfund site. Unfortunately, the pollution has already spread as far as northeastern Oregon, where a landfill has received arsenic-rich soil dug up from near the Tacoma smelter. Final cleaning and monitoring at Holden should be finished in 2020.

Geologists think that the metals in the veins at Holden were originally in ocean-floor basalt and sedimentary rocks of the Chelan terrane, probably the emanations from an oceanic ridge (Chapter 2). Similar deposits are found at submarine hot spring vents today.

131

RAILROAD CREEK VALLEY

A wonderful U-shaped glacial valley

A stop along the trail to Hart Lake above Holden is a good place to view the evidence of glacial erosion. The U-shape of the valley indicates that it was cut by a glacier (Figure 36). Because a glacier is wide and massive, it acts like a broad curved rasp, cutting a wide trench with nearly vertical walls. On the other hand, a mountain stream erodes into the rock like a saw (Chapter 7). In time the vertical walls in both kinds of valley break down and begin to fall back into more gentle slope. The stream valley becomes a V, and eventually the glacial valley, a flat-bottomed V, and given enough time, a broad, open V. The recency of the glacier's work here is proven by the valley's robust U shape.

132

HART LAKE

Intrusive breccia

Just above Hart Lake, the Railroad Creek Trail crosses a smooth open slope, a symmetrical fan of debris dropped by the stream that comes cascading down from the Isella Glacier far above. The cliffs just west of the stream are made of a dense, black igneous rock that, like lava, cooled quickly. A close look at weathered surfaces shows that angular black fragments are in a cement of the same once-molten material which intruded a crack in the crust and forms this dike. Evidently the molten rock had begun to solidify when renewed movements of more molten material broke up the chilled portion and engulfed the pieces.

133

CROWN POINT FALLS OVERLOOK

An old mine adit

On a rocky ledge overlooking Crown Point Falls, the trail passes a dripping entrance to a prospector's adit—that

Figure 103. Lyman Lake with end moraine beyond and Lyman Glacier behind the moraine.

is, a horizontal passageway. On the cliffs just south of Crown Point Falls more adits may be seen above telltale piles of relatively fresh, broken rock. These adits were driven in about 1901 along veins of quartz that contain silvery, metallic flakes of molybdenite. This exceedingly soft sulfide mineral, which can be scratched with a fingernail, is an ore of the metal molybdenum, which is used to make special steel alloys.

134

LYMAN LAKE

End moraine

After about a 45-minute hike south of the Railroad Creek Trail, hikers bound for Spider Pass reach the site of an old snow-survey cabin and an end moraine of the Lyman Glacier (Figure 103). An end moraine forms at the lower end of a glacier, where the melting is just compensated for by ice flowing down from the upper end. The position of the snout remains stationary, but the flowing ice constantly delivers debris to the terminus as if from a conveyor belt (Chapter 7). Above the end moraine at Lyman Lake, an expanse of barren gravel is laced with milky streams from Lyman Glacier and dotted with shallow tarns. Below it a thin film of alpine flora has already spread over similar features. Here is a graphic example of a rapidly changing natural environment.

135

SPIDER PASS

Colorful moraine and black crystals

Narrow, rocky Spider Pass (just off the map, about two hours and 1,500 feet of climbing from the Railroad Creek Trail) lies above the defile that shelters Spider Glacier. From the pass, or north of it, look toward Lyman Lake to see how the rust-colored gneiss blocks carried by Lyman Glacier from Chiwawa Mountain lie in a moraine against the white granite on the east.

At the pass, look for small nests or large masses of radiating black needles of tourmaline in the brown biotite gneiss (Figure 104). Flawless tourmaline crystals, when cut and polished, are semiprecious gemstones. The granodiorite of the Cloudy Pass

Figure 104.

batholith exposed near Lyman Lake extends south here under a roof of much-cooked Swakane Biotite Gneiss, and as the once-molten rock of the batholith cooled, boron-rich emanations entered this fractured roof and deposited the tourmaline (Geologic Notes 26, 34, 64, 115).

136

MARTIN RIDGE

Metamorphosed mud

Hikers on Martin Ridge above Holden have a good opportunity to examine some rocks of the Chelan Mountains terrane. Much of the schist and gneiss on Martin Ridge was once sedimentary and volcanic rock derived from a volcanic arc. The thin, brittle flags of red schist along the crest of Martin Ridge are derived from shale. The red color is iron oxide (rust) from the weathering of iron minerals, particularly tiny, brass-colored cubes of pyrite (iron sulfide), in the rock. East of this red schist is a thick, conspicuous layer of light-colored gneiss. (To reach Martin Ridge, see Crowder and Tabor, 1968, and Beckey, 1989.)

East Side Approaches

METHOW RIVER DRAINAGE

ROADS AND TRAILS REACHED FROM THE METHOW RIVER

137

SOUTH CREEK TRAIL

Remnants of an old ocean

From the end of the west-side Twisp River Road, the trail crosses South Creek and then begins climbing along the foot of the north wall of the South Creek valley. The trail through the woods passes blocks of contorted, stripy quartzite that have fallen from the cliffs above (about one mile and 400 feet of climbing from the trailhead on the Twisp River Road). These are metamorphosed and deformed ribbon chert that probably was deposited in an old ocean about 200–300 million years ago (Triassic or Permian). Metamorphosed basalt, found with the chert in the outcrop above, was the ocean floor. The chert and ocean-floor basalt were buried and metamorphosed sometime before about 95 million years ago (Cretaceous), and then intruded and cooked 90 million years ago by the Black Peak batholith (Geologic Notes 15, 46), which is exposed just north of here. Geologists assign these ocean-born rocks (Chapter 3) to the Chelan Mountains terrane. In this area they are known as the Twisp Valley Schist.

138

EARLY WINTERS

Magmatic gas and the Golden Horn granite

Roadcuts along Highway 20, about 11.9 miles west of the Mazama turn-off and 4.8 miles east of Washington Pass, expose light-colored granite of the Golden Horn batholith. The small (mostly one-quarter-inch diameter or less), irregular, open cavities in the rock are gas pockets, which formed when the rock was still a melt. They show that the magma was rich in water and boiled as it crystallized. Such boiling is most commonly recorded in rocks that crystallized within a few miles of the Earth's surface.

In some rocks here, a white feldspar can be seen forming a mantle or rim around rounded pinkish feldspar. Geologists call this *rapakivi granite,* a name coined in 1735 for similar rocks in Finland. The name is derived from a Finnish word meaning rotten stone, because the Finnish rapakivi granites weather to a coarse gravelly material. Golden Horn granites also weather to coarse sand in some places, as do other kinds of granitic rocks (Geologic Note 45). But rapakivi granite now refers to the texture of the white-rimmed pink feldspar, not the weathering. Geologic jargon, like many jargons, is not always logical.

Golden Horn magma invaded the surrounding rocks about 48 million years ago. The region of the pass here was then about 2 miles down within the Earth. A lot of rock has been uplifted and eroded away since that time, but nowhere near as much as has been removed from some of the region to the west, in the Metamorphic Core Domain. The granite of the batholith is much different chemically from the 75- to 90-million-year-old plutons typical of the Skagit Gneiss Complex and the arc-root plutons of the Cascade Volcanic Arc, which are mostly tonalites or granodiorites (Geologic Note 95). The batholith intruded during the Eocene extensional event, when the forces on the crust were changing. Although geologists don't know exactly why the Golden Horn batholith is chemically different from the arc plutons, perhaps these new crustal forces caused the Golden Horn magma to evolve differently than rocks melted during more typical subduction.

HARTS PASS ROAD (USFS ROAD 5400) AND BEYOND

139
LOST RIVER

Cutting the gorge

The Harts Pass Road drops off a broad terrace above the Methow River (about 6.6 miles from Mazama and just off the map to the east) to cross the Lost River, a waterway that is perhaps one of the more dramatic testimonies to the rearrangement of North Cascade scenery by Pleistocene glaciation. Upriver from the road crossing, the Lost River flows along the bottom of a deep gorge (Figure 105). Some views of this gorge can be had by hiking up the 3.6 miles of trail in the lower end. The canyon walls soar 3,000 to 4,000 feet above the rushing river water. Where the trail leaves the Lost River to climb out of the canyon, the trailless gorge continues for another 10 to 11 miles before the valley bottom widens, and the valley walls fall back into a profile more typical for the region. The pass at the head of the Lost River, where the drainage reverses and flows north into Canada, is almost unnoticeable in the flat valley. In a manner analogous to that which created the Skagit River gorge (Geologic Note 5), the spectacular Lost River canyon was cut by meltwater from the Cordilleran Ice Sheet as it spilled south over a high divide. The rushing meltwater cut a new canyon which, when the ice was gone, remained the drainage for an enlarged area.

We speculate that this great river from the Cordilleran ice carried coarse debris to the Methow and dumped it where the canyon widens, filling the Methow canyon and forming a broad valley floor. A huge amount of water can flow through these coarse, permeable gravel deposits. In late summer, they carry all of the Methow River so that the surface flow disappears altogether. The Methow appears to dry up, though it still flows below, in the gravels of its bed.

140
HARTS PASS ROAD

Red soil reveals the rocks

By the time the Harts Pass Road reaches the first switchback (about 9 miles from Mazama), the soil and surface of the road have become distinctly red. This color derives from the iron oxide mineral hematite, which is abundant in the "red-beds" underlying the hillside here. The red rocks are conglomerate and shale, rich in volcanic debris and deposited about 90 million years ago (Cretaceous). The late Julian Barksdale, Professor of Geology at the University of Washington, called these rocks the Ventura Member (of the Midnight Peak Formation). They represent some of the first sediments washed down from growing volcanoes to the east. The hematite is a sort of rust from iron-rich minerals in the volcanic rocks. Geologists argue about why some rocks rust red and others brown, but most agree that red-beds do not commonly form in the ocean. These are terrestrial sediments deposited by rivers and streams.

141
DEADHORSE POINT

Hard rocks make hard places

Just beyond Deadhorse Point (12 miles from Mazama), drivers may wish to pull over and park at a wide spot,

Figure 105. Trailless Lost River Gorge was eroded by a glacial-melt river of the Cordilleran Ice Sheet. Less steep slopes above the cliffs of the gorge represent the sides of the old, preglacial valley.

then walk back to better comprehend the difficulty of building this road in 1903 and why pack horses did not survive falling off it. Drivers from the midcontinent may think that the road here seems only marginally improved since it was built. The sandstone, siltstone, and shale of the Pasayten Group traversed by the road in this canyon are rarely strong enough to hold up such steep hillsides, but here they are reinforced by numerous near-vertical dikes of rhyolite and basalt. These dikes filled cracks extending out from the Golden Horn batholith (Geologic Notes 16, 138) and related plutons, which were emplaced during the Eocene extensional event, about 50 million years ago (Chapter 5).

142

SLATE PEAK

Viewing rocks of the Methow Domain

Few places give visitors such a splendid view of an entire geologic province as does the summit of Slate Peak. Leaving their cars at the gate on the Slate Peak Road (21.2 miles from Mazama), visitors have to walk only

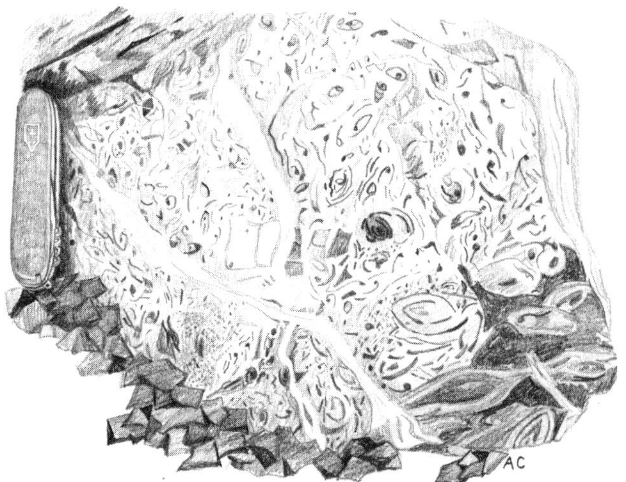

Figure 106. Fossil snails in outcrop along the Slate Peak Road.

200 yards up the road to the summit platform, once the site of a Defensive Early Warning radar station and now capped by a fire lookout tower. To the west on the skyline is Jack Mountain, held up by the Hozomeen terrane, once the ocean floor on which the sedimentary rocks in the foreground were deposited. Beginning about 105 million years ago, the metamorphosed ocean-floor basalt was shoved up and east, on a thrust fault, over the ocean sediments. To the east the rugged, colorful cliffs of Devils Peak and Robinson Mountain are mostly Cretaceous volcanic rocks of the Pasayten Group, capping the immense pile of sedimentary rocks that filled the Methow Ocean. Bedding can be seen on many of the surrounding ridges (see Plate 3B). The disparate inclinations of beds from one ridge to another reflect folds and faults in the rocks (see "Breaking and Folding the Rocks of the Methow Domain" in Chapter 3).

The summit of Slate Peak itself is eroded from river and delta deposits of the Pasayten Group. Back down the road, just before the well-exposed rock in the roadcut on the east gives out to open ridge, and a short distance above the gate at the parking area, is dark siltstone, crowded with fossil snails (Figure 106). These snails flourished about 95 million years ago along a shallow shore of the Methow Ocean, where rivers and streams of the Pasayten Group entered the sea. These encroaching deltas of the Pasayten Group eventually finished off the dying Methow Ocean (Figure 22).

The light-colored white-to-orange rocks along here are volcanic dikes, which are easily mistaken for sandstone beds. Small rectangular white flecks of feldspar and rounded, glassy quartz grains are visible in an otherwise fine-grained matrix. Some 40 million years after the snails had died, these dikes filled cracks in the sedimentary rocks during the Eocene extensional event (Chapter 5).

143

CHANCELLOR

A long-gone mining center

Visitors reaching ruins of Chancellor by way of the Canyon Creek Trail or the old road from Harts Pass may want to contemplate the brief budding of civilization in the area. Prospectors found gold in the rocks and in the stream gravels in this area as early as 1877. In 1903 a trail to Chancellor over Harts Pass was widened to accommodate wagons, and by 1905 a hydroelectric plant was operating to run mines and mills in the surrounding area. Lost in the woods of Slate Creek, near the confluence of Bonita Creek, are the remains of a once busy mining camp, complete with store and tavern. Mines above Barron and in Canyon Creek and its tributaries produced gold and silver which was milled locally, while placer miners panned and engaged in hydraulic mining along Canyon and Ruby Creeks (Geologic Note 43).

By the late 1930s some 1.6 million dollars in gold and silver had been removed from the Slate Creek district, most with great effort. The Azurite Mine and mines in the Barron area (Geologic Notes 145, 148) produced the most ore. The metals occur in quartz veins cutting the sedimentary rocks of the Methow Ocean and probably came from nearby stitching plutons that invaded the sedi-

Figure 107. Shales of the Pasayten Group (Virginian Ridge Formation) overlying sandstone of the Methow Ocean (Harts Pass Formation) along unconformity (looking northeast from the Pacific Crest Trail; Geologic Note 146).

mentary rocks about 90 million years ago (Cretaceous). Although the rocks still contain ore minerals, the rigors of winter and changing economics have closed down almost all the mines. The forest has taken over once again.

144

CADY PASS

Gravels swept into the Methow Ocean

The views at Cady Pass (accessible along the old road to the Azurite Mine, about 5 miles from the Slate Creek Road) are magnificent. But to see some interesting rocks, hike south across the meadows and climb up on glacially polished slabs of conglomerate. These conglomerates may have formed in the deep water of submarine canyons as density flows that swept out onto the deep ocean floor (Chapter 3). The conglomerate contains widely spaced strings of pebbles aligned in a sandstone matrix. About 100 million years ago (Cretaceous) storms may have stirred the coarse material into a density slurry in shallow waters near shore. The slurry, loaded with pebbles, then flowed down the submarine canyon.

145

AZURITE MINE

A long haul for gold

The seldom-visited Azurite Mine lies at the bottom of an impressive canyon. The rugged walls of Majestic Mountain and Mount Ballard rise 4,000 feet above the tired mine buildings and yellow-orange tailing. The Mill Creek valley bottom is rich in the artifacts of hard work and the hope of riches. The Azurite Mine was opened in 1915. Pack strings carried gold and silver ore out to the Methow Valley until 1930, when the road was completed. Mining ceased in 1939. More recently, would-be miners have focused on the gold still left in the tailing.

146

SOUTH OF HARTS PASS

View of an unconformity

Just a short hike south from Harts Pass on the Pacific Crest Trail, near the Black Bear prospect, before the trail swings around a ridge to go west, look east above Rattlesnake Creek to view an unconformity between deep-sea sandstone of the Methow Ocean—the Harts Pass Formation—and overlying deltaic and stream deposits of the Pasayten Group (Figure 107). The viewpoint is located about one-half mile south of the auxiliary trailhead near Meadows Campground.

The trail here was hacked out of steeply inclined, thick beds of Harts Pass sandstone. To the east, similar thick beds of sandstone make bold outcrops above the Harts Pass Road. The strata were uplifted, slightly tilted, and eroded before being covered by flood-plain shales and river gravels of the Virginian Ridge Formation above the unconformity. The shaley strata make smoother slopes above the Harts Pass outcrops, interrupted here and there by more prominent beds of sandstone and conglomerate deposited in the channels of the ancient rivers. We view here the final cover that helped eliminate the Methow Ocean. Both deep-sea sandstone and river deposits were intruded about 50 million years later by orange-weathering dikes (Geologic Notes 141, 142).

The Harts Pass strata were uplifted and eroded over a period of only a few million years before streams and rivers covered them with the new deposits. Today, the

Figure 108. Fold in sandstone and shale of the Methow Ocean as viewed from Tatie Peak (looking southwest).

scene is an unconformity in the making, as erosion wears down the tilted strata of both formations. Will future geologists find yet more sediments deposited on the eroded stumps of the North Cascades and, after more erosion, view two unconformities in this spot?

147

TATIE PEAK

View of small fold

Hikers on the Pacific Crest Trail south of Harts Pass are treated to a nice view of folded rocks. At about 1 mile south of the trailhead, they can look southwest to a broad peak between Grasshopper Pass and Tatie Peak to see a fold in the strata of the Harts Pass Formation (Figure 108). The bold ribs of marine sandstone stand out between softer shale beds, and their curving trace on the mountainside reveals the fold. Some 105 million years ago, sand and mud collected in the Methow Ocean. Ten or 15 million years later, the sand and shale, by then converted to rock, were folded as tectonic plate movements squeezed the Methow Ocean. Now, some 90 million years later, the fold has been uplifted and etched out by erosion on the mountainside.

148

BARRON

The old scars of gold mining

Hikers on the Pacific Crest Trail north of Slate Peak can hardly miss viewing the scars and scabs of mining endeavors across Bonita Creek to the west. During the height of the gold rush to the Slate Creek District in the late 1800s and early 1900s, more than two thousand claims were staked on the hillsides above Bonita Creek. Remnants of mill and buildings remain at the mining camp of Barron (Figure 109), perhaps visible from the trail approaching Windy Pass. Many of the claims are still valid under the 1872 mining laws. The owners need only to perform a hundred dollars worth of "assessment" work yearly to maintain a claim.

149

HOLMAN PASS

River over the ridgecrest

Backpackers slogging through the forest of low and broad Holman Pass may relieve the monotony by wandering west into upper Canyon Creek and looking at rounded bedrock outcrops, which are somewhat unusual in such a low and wooded reach. Where is the cover of weathered rock and glacial debris that usually hides such outcrops from rock-minded hikers?

Holman Pass is the lowest pass separating the north-flowing Pasayten River drainage from the south-flowing Canyon Creek–Skagit River drainage. When the advancing (and retreating) Cordilleran Ice Sheet filled the Pasayten valleys to the north, a great lake formed in the middle reaches of the Pasayten River drainage. At certain ice levels, this lake drained to the southwest, through Holman Pass, into Canyon Creek. The torrent of glacial meltwater stripped off all the loose debris on the west

Figure 109. Barron gold camp about 1895 (from Moen, 1976).

(downstream) side of Holman Pass and farther downstream helped excavate the impressive gorge of Canyon Creek (Geologic Note 43).

150

WOODY PASS

Cobble conglomerate

Along the Pacific Crest Trail at, and south of, Woody Pass, hikers can admire extensive glacially polished outcrops of cobble conglomerate in the rocks of the Methow Ocean. The cobbles are of a variety of rock types, including abundant black-speckled hornblende-bearing granitic rocks (Plate 5B). Such conglomerates can be beach or river deposits, but here they are sandwiched between sandstone and siltstone whose features indicate deposition from density flows (see "Sand in the Sea: How Does It Move" in Chapter 3). Geologists conclude that these conglomerates were deposited by the densest part of these density flows in immense channels in deep-water submarine fans that were built into the Methow Ocean about 105 million years ago (Geologic Notes 38, 144).

Geologists have measured indicators of current-flow in these rocks and find that the density flows moved from east to west. The submarine fans must have been fed with debris eroded from granitic rocks somewhere to the east. But what granitic rocks were they? Prior to the birth of the concept of far-traveled tectonic terranes, geologists would have assumed these cobbles came from the granitic rocks that now crop out in the Okanogan Range some 20 miles east of here, on the other side of the Pasayten Fault. With the suggestion from paleomagnetic studies (Geologic Note 15) that some terranes of the North Cascades traveled thousands of miles from their origins to the south, we have to consider that the Methow Ocean may have been thousands of miles to the south also. Maybe these granitic cobbles came from granitic rocks then somewhere in southern Mexico (Chapter 5).

151

THE BOUNDARY TRAIL

Chert pebble conglomerate

The Boundary Trail from Castle Pass provides grand scenery to the south and outcrop after outcrop of thin-bedded sandstone and siltstone along the trail. All the beds dip to the west and have many features showing that younger and younger beds lie to the west. Some beds have a distinct change in the size of sand grains, from coarse sand at the base of beds (on the east) to fine sand or silt at the top (on the west). This grading develops as a density flow on the ocean bottom comes to rest and its load of sediment settles out. In addition, some sandstone beds have distorted, lumpy bottoms formed as the sand compacted into underlying mud (see "Sand in the Sea: How Does It Move" in Chapter 3).

The beds here were deposited in submarine fans of the Methow Ocean. At the head of Two Buttes Creek (about 3 miles from Castle Pass), the trail crosses beds of conglomerate, with chert pebbles eroded from the upthrust ocean floor that became the Hozomeen terrane. These conglomerates also contain a grading of scattered pebbles in sandstone beds, with the larger pebbles at the bottoms. Fossils from related rocks farther south indicate these west-derived beds are between 100 and 105 million years old (Cretaceous). They overlie rocks to the southeast along the Pacific Crest Trail (Geologic Note 150), which have features showing that the sediment there came into the ocean from the east. In other words, the ocean was filling in from both sides.

East Side Approaches

PASAYTEN RIVER DRAINAGE AND BEYOND

152

WEST FORK OF THE PASAYTEN RIVER

Monster landslide

Hikers along the lower West Fork of the Pasayten may well wonder when they will leave the winding, sometimes brushy trail and reach better views, or even camp and dinner. This is a good place to think about a catastrophic collapse of Buckskin Point to the east. About 6 miles south of Pasayten Airfield and 8 miles north of the Slate Peak Road, the West Fork Pasayten Trail traverses the debris from this ancient landslide for about a mile and a half (Figure 110). From favorable places along the trail, hikers can look east to see irregular humps and bumps of the landslide deposits. Glacial geologist Richard Waitt of the U.S. Geological Survey determined that the landslide came down on top of a tongue of the Cordilleran Ice Sheet and was transported a short distance south by the flowing ice. Eventually, the well-traveled debris was slowly let down onto the ground as the ice melted away beneath it.

153

DEAD LAKE

Dead clams

Searching the talus around Dead Lake will usually turn up fossils (mostly holes left in the rock after the fossil

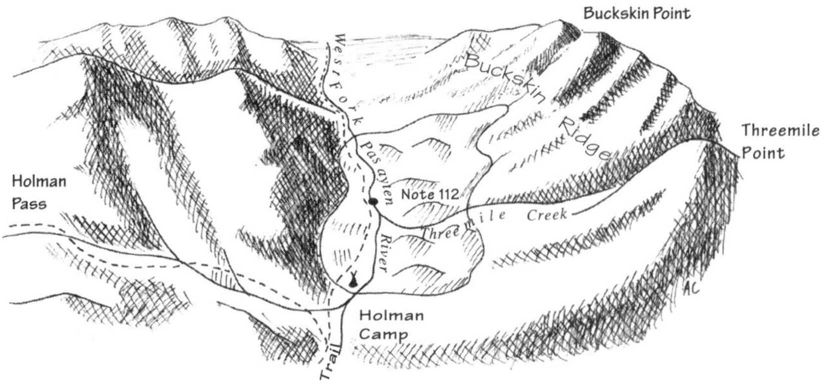

Figure 110. Ice-age landslide in the valley of the West Fork of the Pasayten River (extent of landslide after Waitt, 1979)

Figure 111. Fold on the north side of Mount Rolo. Diagrammatic cross section at top shows how fold develops and breaks along thrust fault.

has dissolved out) of *Trigonia*, a rough-shelled clam that lived some 110 million years ago (Cretaceous). The fossils appear to have been preserved in storm deposits, coarse-grained sandstone beds deposited after storm waves stirred a shallow sea. The process is like that of shaking a bottle of water, sand, gravel, shells, and mud, then letting the contents settle. The gravel and shells will settle first, making a coarse-grained deposit that underlies the more slowly settling sand and mud. The shallow water indicated by the Dead Lake deposit and others like it contrasts with the earlier deep water of the Methow Ocean, as indicated by features in rocks older than these. Apparently, the ocean was filling up, except that rocks younger than these (Geologic Notes 150, 151) indicate the ocean was deep again. The likely explanation of this anomalous later deepening is that out to the west, geologically soon after these clams were deposited, tectonic forces thrust the ocean floor upward and eastward over the sediments. The weight of this basaltic thrust plate depressed the ocean basin, causing it to deepen again before filling up one last time.

154

MOUNT ROLO

Folding and thrusting

Cresting the ridge south of Freds Lake, hikers come face to face with a large fold on the north face of Mount Rolo, some three-quarters of a mile to the south (Figure 111). The folded rocks, which consist of thick beds of sandstone and intervening thin beds of shale, are density-flow deposits that filled the Methow Ocean about

105 million years ago (Cretaceous). The steep beds on the east and shallowly dipping beds on the west suggest that the fold was enhanced by a push from the west.

Hikers can admire the view and, when refreshed, continue on the trail, but geologists are obliged to descend south, down several hundred feet, into the headwaters of Eureka Creek, where they learn to their surprise that they are here surrounded by river-deposited sandstone of the Pasayten Group (about 95 million years old). These rocks are younger than the rocks of the Methow Ocean that occur above on the ridgecrest and Mount Rolo (Chapter 3; see Plate 2). Features in the sandstone beds both here and on the ridgecrest show the beds to be right-side up, which would mean that rocks on the ridgecrest should be younger. This is most unusual! A large thrust fault—the Chuchuwanteen Thrust—has brought up older marine rocks of the Methow Ocean and shoved them eastward over younger sandstone of the Pasayten Group.

Glossary

alluvium. Sand, gravel, and silt deposited by rivers and streams in a valley bottom. An **alluvial fan** forms a half-cone-shaped accumulation of sediment where a side stream debouches onto a flatter-floored valley.

amphibole. A family of silicate minerals forming prismatic or needlelike crystals. Amphibole minerals generally contain iron, magnesium, calcium, and aluminum in varying amounts, along with water. **Hornblende** always has aluminum and is a most common dark green-to-black variety of amphibole, forming in many igneous and metamorphic rocks. **Actinolite** has no aluminum and is needle-shaped and light green. **Blue amphibole** contains sodium and, of course, is bluish in color.

amphibolite. Generally a metamorphic rock, mostly made of hornblende and plagioclase.

andesite. Fine-grained, generally dark-colored, igneous volcanic rock with more silica than basalt. Commonly with visible crystals of plagioclase feldspar. Generally occurs in lava flows, but also as dikes. The most common rock in volcanic arcs.

arc rocks. Volcanic arc rocks. See *volcanic arc.*

argillite. Unusually hard, fine-grained sedimentary rocks, such as shale, mudstone, siltstone, and claystone. Commonly black.

banded gneiss. See *gneiss.*

basalt. Very fine-grained, generally black, volcanic igneous rock relatively rich in iron, magnesium, and calcium. Generally occurs in lava flows, but also as dikes.

base level. The elevation at which a stream or river can erode no more, usually sea level. May be a temporary or local base level, such as a lake or very hard rock, which significantly slows down erosion for the reach immediately upstream.

batholith. Very large mass of slowly cooled, intrusive igneous rock such as granite. Geologists say it must be at least 50 square miles in exposed area. See *pluton, stock.*

bedding. Sedimentary layers in a rock. Beds are distinguished from each other by grain size and composition. Subtle changes, such as beds richer in iron oxide, help distinguish bedding. Most beds are deposited essentially horizontally.

blueschist. Metamorphic rock rich in blue amphibole. See *amphibole.*

breccia. Rock made up of angular fragments of other rocks. **Volcanic breccia** is made of volcanic rock fragments generally blown from a volcano or eroded from it. **Fault breccia** is made by breaking and grinding rocks along a fault.

calcite. Mineral made of calcium carbonate ($CaCO_3$). Generally white, easily scratched with knife. Most seashells are calcite or related minerals. The lime of limestone.

caldera. Large, generally circular, fault-bounded depression caused by the withdrawal of magma from below a volcano or volcanoes. Commonly, the magma erupts explosively, as from a giant volcano, and falls back to Earth as volcanic ash.

chert. Sedimentary rock made of extremely fine-grained quartz. Usually made of millions of globular siliceous skeletons of tiny marine plankton called radiolarians. Black chert is called flint.

chlorite. Family of platy silicate minerals containing various amounts of magnesium, iron, aluminum, water, and small amounts of other elements. Some mineralogists include chlorites in the mica family because the crystals form small flakes. Commonly green.

clay. Family of silicate minerals containing various amounts of aluminum, potassium, and sodium as well as water. Generally form platy crystals too small to be seen even with a microscope. Can form at room temperature and hence is a common product of rock weathering, especially of rock containing lots of feldspar. The term *clay* is also used to refer to very, very fine sedimentary grains whether or not they really are made of clay minerals.

conglomerate. Sedimentary rock made of rounded rock fragments, such as pebbles, cobbles, and boulders. To qualify as a conglomerate, some of the consituent pebbles must be at least 2 mm (about 1⁄13th of an inch) across.

Cordilleran Ice Sheet. Ice cap that grew in western North America during the Pleistocene Epoch, beginning in Canada and covering much of British Columbia, Alaska, and the northernmost states of the western United States.

delta. Sedimentary deposit formed where a river enters the ocean or a lake and drops its load of sand and silt. Generally fan-shaped, a delta is fed by many meandering distributor streams, and builds outward from the shore into the lake or ocean.

diapir. Forceful upward, pistonlike intrusion of a rock mass into overlying rock. In the case of an igneous diapir, the intruding rock may be magma or a crystal-rich mush, either of which is lighter (less dense) than the surrounding rock. It works its way by upwelling in the core of the pistonlike mass. Imagine the upwelling of thick soup in a kettle over a small hot burner.

dike. Tabular body of igneous rock formed where magma fills a crack in preexisting rock. A **sill** is a dike that parallels sedimentary bedding or metamorphic foliation.

diorite. Intrusive igneous rock made of plagioclase feldspar and amphibole and/or pyroxene. Similar to gabbro only not so dark and containing, chemically, less iron and magnesium. See Figure 112.

epidote. Family of silicate minerals mostly containing calcium, aluminum, iron, and magnesium, along with water. Apple green to straw yellow, epidote generally forms very small, stubby, prismatic crystals. Most common in metamorphic rocks, but some forms occur in igneous plutons that crystallize very deep in the crust.

extension. In geology, the process of stretching the Earth's crust. Usually cracks (faults) form, and some blocks sink, forming sedimentary basins.

fault. Crack in the Earth's crust where the rocks on one side of the break move relative to the rocks on the other. Abrupt movements on faults cause earthquakes. Where the crack is roughly vertical the rocks may move up or down or sideways or in some combination. If the fault is inclined at a low angle to the Earth's surface and rock on one side of the fault moves up and over rock on the other side, it is a **thrust fault**. On maps and cross-sectional diagrams, geologists and the authors of this book commonly show the relative

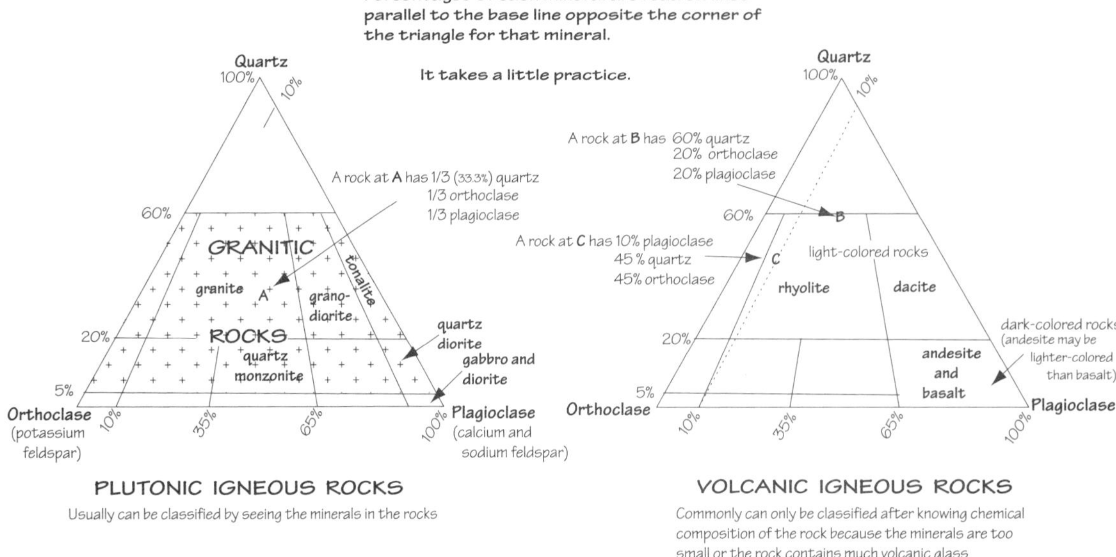

Figure 112. Classification of igneous rocks by content of light-colored minerals (shown at the corners of the triangular diagrams). Dark minerals like biotite and hornblende are not counted in this scheme. All rocks in patterned field of the plutonic triangle can be called granitic rocks. Only selected rock names are shown. See also Figure 86.

---

motion of faults with half-arrows drawn on either side of the fault.

feldspar. Family of silicate minerals containing varying amounts of aluminum, potassium, sodium, and calcium. **Potassium feldspars** contain considerable potassium. **Plagioclase feldspars** contain considerable sodium and calcium. Feldspar crystals are stubby prisms, generally white (plagioclase) or pink (potassium feldspar) and with the look of porcelain.

fission tracks. Microscopic tunnels made in crystals by escaping nuclear particles emitted by radioactive elements. Most commonly studied are fission tracks in zircon crystals made by the radioactive decay of uranium, present as an impurity.

foliation. Parallel arrangement of minerals, especially platy minerals such as micas, in a rock, so as to give it a foliated look, like pages in a book or a pile of leaves arranged to lie flat. Foliated rocks tend to break along the foliation and form slabs. Mostly in metamorphic rocks.

gabbro. An igneous rock made of calcium-rich plagioclase, with amphibole or pyroxene. A dark coarse-grained rock, chemically equivalent to basalt. See Figures 86 and 112.

garnet. Family of silicate minerals containing varying amounts of iron, magnesium, calcium, and aluminum. Schist or gneiss commonly have garnets, which look like tiny, glassy red spheres, but are really dodecahedrons.

gneiss. A light-colored metamorphic rock with foliation or lineation but with the minerals *and the look* of a granitic rock. Feldspar is prominent. **Banded gneiss** is made of alternating layers of darker schist or amphibolite and lighter-colored granitic gneiss. The bands are really layers seen on edge.

graben. A block of rock that has dropped down between two faults. Commonly the block is covered with sedimentary rocks, which filled the depression. Later, the sedimentary filling erodes more readily to create a lowland between ridges of more resistant rocks (for example, the Chiwaukum Graben between the metamorphic rocks of the Entiat and Chiwaukum Ranges). Usually the graben is low because of differential erosion, not because the faults are still down-dropping the block.

granite. A coarse-grained igneous rock with considerable potassium feldspar, as well as quartz and subordinate plagioclase feldspar in visible crystals. Usually with biotite, but also may have hornblende. See Figures 86 and 112.

granitic rocks. A general term for an igneous rock that looks like granite but may range in composition from quartz diorite to granite. All granitic rocks are light colored, with feldspar and quartz visible in a hand specimen. See Figures 86 and 112.

granodiorite. A coarse-grained igneous rock with more plagioclase than potassium feldspar. Otherwise like granite. See Figures 86 and 112.

greenstone. A metamorphic rock derived from basalt or chemically equivalent rock such as gabbro. Greenstones contain sodium-rich plagioclase feldspar, chlorite, and epidote, as well as quartz. The chlorite and epidote make greenstones green.

hanging valley. A tributary valley whose floor is noticeably higher than that of the main valley into which it opens. Generally caused by the more effective erosion of a trunk-valley glacier than its smaller tributary glacier. The step between the two floor levels supports a waterfall or rapids. Commonly, the tributary stream in the hanging valley cuts into, or notches, the step, so that the steep-walled canyon of the tributary where it joins the main trunk stream is most noticeable.

hornblende. See *amphibole*

hornblende schist. A schist rich in hornblende. Generally with abundant plagioclase feldspar as well. Grades into amphibolite.

hornfels. A dark, very fine-grained metamorphic rock produced by the recrystallization of a fine-grained rock by heat from a nearby igneous intrusion. From the German, meaning horn rock.

igneous rocks. Rocks formed by crystallization of molten rock. **Plutonic igneous rocks** form at depth within the Earth and have crystallized slowly. Their crystals are coarse. **Volcanic igneous rocks** form at or near the Earth's surface and crystallize rapidly. Their crystals are fine.

intrusion. Usually an igneous rock that is an intruder into preexisting rocks. Dikes, sills, and batholiths are intrusions.

intrusive. Refers to an igneous rock that has intruded preexisting rocks. Contrast with volcanic rock on the Earth's surface, which is called **extrusive** igneous rock.

isotope. Many elements are made up of atoms with slightly different numbers of subatomic particles (neutrons) in their nucleus. **Radioactive isotopes** are unstable and shed subatomic particles over time until they become stable. For instance, unstable isotopes of uranium break down to become lead.

joint. A major crack in rock. Joints commonly form in parallel sets.

landslide. Any piece of the Earth's crust that has slid downhill. May be made of rock, soil, alluvium, votes, etc.

lawsonite. A metamorphic mineral that forms only under very high pressure at relatively low temperature. It is a calcium aluminum silicate and usually forms microscopic crystals.

lava. Magma that has flowed out onto the Earth's surface.

limestone. A rock made up of the mineral calcite (calcium carbonate). Commonly formed from the calcium carbonate shells of marine creatures.

limonite. A mineral composed of iron oxides and water. Rust. Very common in many rocks after weathering at the Earth's surface. Imparts brown or yellow colors to many rocks.

lineation. Parallel arrangement of elongate minerals or groups of minerals. To envision lineation, imagine packages of spaghetti or pencils.

lithosphere. Outer shell of the Earth characterized by being rigid compared to deeper plastic material (the **asthenosphere**). The lithosphere is made of the crust and the uppermost part of the mantle.

magma. Melted rock formed deep in the Earth. When magma pours out on the Earth's surface it is called **lava**. A **magma chamber** is a mass of melted rock deep in the earth.

magnetite. Iron oxide mineral (Fe_3O_4). Usually tiny black, metallic crystals. Magnetite will attract a magnet and sometimes, in a rock, a hiker's compass needle.

magnetometer. Device for measuring magnetism.

mantle. Interior part of the Earth surrounding the core and below the crust. Made up of dense, iron- and magnesium-rich (ultramafic) rock such as dunite and peridotite.

marble. Metamorphic rock of calcium carbonate derived from limestone by recrystallization.

mélange. Mixture of rocks formed by tectonic disruption, such as multiple faulting, which brings disparate rock types together. Usually consists of a matrix of weak material, like shale, with hard pieces of exotic rocks, such as gneiss or igneous rocks.

metaconglomerate. Metamorphosed conglomerate.

metamorphic rocks. Changed rocks. Generally rocks with new minerals or textures derived from older rocks by squeezing and/or heating and recrystallization deep in the earth.

mica. Group of silicate minerals composed of varying amounts of aluminum, potassium, magnesium, and iron, as well as water. All micas form flat, platelike crystals, which cleave into smooth flakes. **Biotite** is dark, black or brown mica. **Muscovite** is light-colored or clear mica, sometimes called isinglass when in large, useful flakes.

migmatite. Mixed rock. Migmatites are usually conspicuously mixed light- and dark-colored rocks, generally formed of dark-colored schist and/or gneiss and lighter igneous intrusive dikes and sills intruded during deformation; they appear to have been mushed around while still hot and plastic.

mineral. A naturally occurring element, chemical compound, or limited mixture of chemical compounds. Minerals generally form crystals and have specific physical and chemical properties that can be used to identify them.

moraine. A hill-like pile of rock rubble located on or deposited by a glacier. An **end moraine** forms at the terminus of a glacier. A **terminal moraine** is an end moraine at the farthest advance of the glacier. A **lateral moraine** forms along the sides of a glacier. See Figure 37. See *till.*

mudstone. A very fine-grained sedimentary rock formed from mud.

nonsilicate mineral. A mineral without silicon (Si). See *silicate minerals.*

oceanic rocks. Rocks formed in the deep ocean. Includes sedimentary rocks deposited on the deep ocean floor as well as the basalt of the oceanic crust. Commonly include some slices of the underlying mantle (ultramafic rocks) as well.

olivine. Silicate mineral containing iron and magnesium. A green glassy mineral formed at high temperature. Common in basalt, especially ocean-floor basalt, and in ultramafic rocks. Gem-quality olivine is called **peridote.** Rock made up mostly of olivine is called **peridotite.** Rock made up entirely of olivine is called **dunite.**

orthogneiss. Gneiss formed by squeezing (deformation and usually some recrystallization) of a granitic igneous plutonic rock.

outwash. Glacial outwash is the deposit of sand, silt, and gravel formed downstream from a glacier by meltwater streams and rivers. An **outwash plain** is an extensive, relatively flat area of such deposits.

paleomagnetism. Literally, early magnetism, meaning magnetism formed in a past geologic era. Generally refers to the magnetism of a rock imparted to it by the Earth's magnetic field when the rock formed.

pegmatite. Generally very coarse-grained rocks of feldspar and quartz. Forms sills and dikes, commonly with other minerals, including very rare ones. Crystals are usually over an inch across. In some pegmatites, feldspar crystals are several feet across.

petrology. Study of rocks.

petrophile. Lover of rocks. May be a petrologist, a stone mason, a rock climber.

phyllite. A very fine-grained, foliated metamorphic rock, generally derived from shale or fine-grained sandstone. Phyllites are usually black or dark gray; the foliation is commonly crinkled or wavy. Differs from less recrystallized slate by its sheen, which is produced by barely visible flakes of muscovite (mica).

plankton. Generally tiny animals or plants that live floating in water.

pluton. Plutonic igneous body of rock that crystallized deep in the earth from magma. **Batholiths** are made of one or more plutons. **Stocks** are generally single plutons, but dikes and sills are not normally considered plutons unless they are very big.

porphyry. Igneous rock, usually a dike or sill, with larger, generally conspicuous, early-formed crystals contained within a matrix of much smaller crystals.

precipitate (or **mineral precipitate**). A mineral deposited from a water solution in pores or other openings in rocks. Chemical reaction with the surrounding rock, changes in pressure or temperature, or evaporation can cause a mineral to precipitate out of solution. Quartz veins are common products of mineral precipitation.

pyrite. Iron sulfide mineral (FeS). Forms silvery to brassy metallic cubes or masses. Common in many rocks. Known as fool's gold. Weathered pyrite produces **limonite** (iron oxide with water) that stains rock brown or yellow.

pyroxene. Family of silicate minerals containing iron, magnesium, and calcium in varying amounts. Differs from amphibole family by lack of water in its crystal. The most common variety, **augite,** contains aluminum as well. Generally forms very dark green to black stubby prisms.

quartz. Silicon dioxide (SiO_2). Also called silica. One of the most common minerals in the Earth's crust (and some new-age boutiques). Crystals are clear, glassy six-sided prisms. Commonly in white masses.

quartzite. Hard, somewhat glassy-looking rock made up almost entirely of quartz. Metamorphosed quartz sandstone and chert are quartzites.

radiocarbon age. The age of organic material determined by the amounts of carbon isotopes 12, 13, and 14. The ratio of 12 to 14 is about the same in all living things, but when a plant or animal dies no more carbon is taken on. Carbon 12 and 13 are stable isotopes, and the amounts remain the same even in dead material. Carbon

14 is an isotope that decays radioactively until none is left. Thus, the ratio records the time elapsed since death. Since carbon 14 decays relatively rapidly, the method is only reliable for the last 40,000 years.

radiolarian chert. A rock made up of the spherical siliceous shells of **radiolarians,** which are single-celled planktonic animals (protozoans).

rhyolite. A volcanic rock chemically equivalent to its plutonic counterpart, granite. Usually light-colored, very fine-grained or glassy-looking. May have tiny visible crystals of quartz and/or feldspar dispersed in a glassy white, green, or pink ground mass.

ribbon chert. Chert and shale in thin alternating beds. The beds resemble parallel ribbons stretched over an outcrop.

root of volcano. Plutonic igneous rock formed from magma that crystallized beneath the volcano it once fed.

schist. Conspicuously foliated metamorphic rock usually derived from fine-grained sedimentary rock such as shale. **Mica schist** is rich in mica such as biotite and/or muscovite.

sedimentary rocks. Rocks formed from pieces of preexisting rocks that have accumulated on the Earth's surface. Most consist of small grains once transported by water.

serpentine. A family of silicate minerals rich in magnesium and water, derived from low-temperature metamorphism of the minerals in ultramafic rocks. Rocks made up of serpentine minerals are **serpentinite.** Serpentine minerals are light to dark green, commonly varied in hue, greasy-looking, and slippery to the touch.

shale. Sedimentary rock derived from mud. Commonly finely laminated (bedded). Particles in shale are commonly clay minerals, mixed with tiny grains of quartz eroded from preexisting rocks. **Shaley** means like a shale or having some shale component, as in shaley sandstone.

silicate. Refers to the chemical unit silicon tetroxide, SiO_4 that is the fundamental building block of **silicate minerals.** Silicate minerals make up most rocks we see at the Earth's surface.

siliceous. Generally refers to a rock rich in silica (SiO_2).

sill. See *dike.*

stitching plutons. Plutons of roughly the same age that intruded several tectonic terranes after the terranes were faulted together. The plutons do not really "sew" the terranes together, but they help record when terranes were assembled.

stock. Relatively small globule- or column-shaped body of plutonic igneous rock. Like a batholith only smaller.

stream capture. A process of erosion where one stream erodes headward, diverting some of another stream's drainage into its own channel. Also called **stream piracy.**

subduction. Process of one crustal plate sliding down and below another crustal plate as the two converge. The **subduction zone** is the area between the two plates, somewhat like a giant thrust fault.

submarine fan. Fan- or cone-shaped accumulation of sedimentary debris—sand, gravel, mud—under the ocean along the edge of the land, either a continent or a volcanic arc. Fans may be a few miles to a hundred or so miles across.

talc. Magnesium silicate mineral, with water. Commonly called **soapstone.** Very soft and platy, like mica. Can be easily carved with a knife. Generally in very fine-grained masses.

talus. Pile of rock rubble below a cliff or chute. **Talus slope** is a common usage although it is redundant because the term *talus* actually includes the concept of a slope.

tarn. Small lake left after the retreat of a glacier. May fill a basin formed by a moraine dam or eroded by the glacier into bedrock.

terrace. A level or near-level area of land, generally above a river and separated from it by a steeper slope. A **river terrace** is made by the river at some time in the past when the river flowed at a higher level. A terrace may be made of river deposits such as gravel or sand, or it could be cut by the river on bedrock. A **glacial terrace** or outwash terrace is similar but is formed by a stream or river from a glacier upstream.

terrane. A rock formation or assemblage of rock formations that share a common geologic history. A geologic terrane is distinguished from neighboring terranes by its different history, either in its formation or in its subsequent deformation and/or metamorphism. A terrane must be separated from its neighbors by faults. Formations within one terrane may have different birthplaces (see Figure 19). Formations that originated at roughly the same time as part of the same paleogeographic scene (for example, deposits of a volcanic arc spilling into a deep ocean basin) and that continued to share the same geologic history, such as metamorphism, traveling, or the like, are considered part of the same terrane. An **exotic terrane** is one that has been transported into its present setting from some great distance.

thrust fault. See *fault.*

thrust plate. Slab of rock, generally on the scale of a mountain or greater, bounded by one or more thrust faults.

till. Unsorted, unstratified rock rubble or debris carried on and/or deposited by the ice of a glacier.

tonalite. Intrusive igneous rock made of plagioclase feldspar, quartz, and amphibole or biotite. May be similar to diorite but contains considerable quartz, is not as dark, and chemically has less calcium, iron, and magnesium. See Figures 86 and 112.

tuff. Volcanic rock made up of **volcanic ash** (volcanic rock fragments and mineral crystal fragments that are mostly smaller than ¼ inch across). Volcanic ash is composed of much shattered volcanic rock glass—chilled magma blown into the air and then deposited.

ultramafic rock. Rock very rich in iron and magnesium and with much less silicon and aluminum than most crustal rocks. Igneous varieties are peridotite and dunite. Most come from the Earth's mantle. A common metamorphic variety is serpentinite.

unconformity. The contact between older rocks and younger sedimentary rocks in which at least some erosion has removed some of the older rocks before deposition of the younger. An unconformity represents a time gap in the record. An **angular unconformity**—where there is an angle between the truncated older beds and the overlying younger beds—shows that the older rocks have been deformed and eroded before the younger sedimentary rocks were deposited.

vein. Tabular rock filling of a generally small crack such as a **quartz vein.** A product of chemical precipitation from a watery solution, in contrast to a dike crystallized from magma, although gradations exist.

volcanic arc. Arcuate chain of volcanoes formed above a subducting plate. The arc forms where the descending plate becomes hot enough to release water and gases that rise into the overlying mantle and cause it to melt. **Arc rocks** are mostly volcanic rocks from the volcanoes and sedimentary rocks made up of eroded debris from the volcanoes. Melted rock in the deeper plumbing of the arc may crystallize at depth to become an **arc-root pluton.**

volcanic rocks. Rocks formed at or very near the Earth's surface by the solidification of magma. Volcanoes produce volcanic rock.

weathering. Process of rock alteration and degradation on the Earth's surface. Caused by chemical changes induced by water and organic acids of plants, as well as by mechanical processes such as water in cracks freezing and expanding, or temperature changes that expand and shrink individual minerals enough to break them apart.

zircon. Mineral of zirconium, silicon, and oxygen (zirconium silicate). Generally glassy-looking, microscopic four-sided prisms. Most commonly formed in igneous rocks.

Reading and References

Unless otherwise stated, most of these references are highly technical. References preceded by an asterisk (*) may be of particular interest to the readers of this guide. References significant to more than one chapter or section are fully cited at the first appropriate chapter or note only; later citations are cross-referenced back to the original citation. References were last updated for the fourth printing (2010).

PART I: NORTH CASCADES GEOLOGY

CHAPTER 1. WORLD CLASS AND CLOSE TO HOME

*Daly, R. A., 1912. Geology of the North American Cordillera at the forty-ninth parallel. Geological Survey of Canada Memoir, v. 38, 840 p. *Less technical than most papers. Interesting, rambling account revealing early development of geological ideas.*

Gibbs, George, 1874. Physical geography of the northwestern boundary of the United States. *Journal of American Geography Society, New York,* v. 4, pp. 298–392. *An interesting travelogue.*

Mierendorf, R. R., 1993. Chert procurement in the upper Skagit River valley of the Northern Cascade Range, Ross Lake National Recreation Area, Washington. National Park Service Technical Report, NPS/PNRNOCA/CRTR-93-001, 115 p.

CHAPTER 2. A GEOLOGY PRIMER FOR THE NORTH CASCADES

Any good geology textbook will help a reader lacking geologic background. Three particularly good ones with great diagrams and photographs (and available in paperback) are:

Press, Frank, and Siever, Raymond, 1994. *Understanding Earth.* W. H. Freeman and Co., New York, 593 p.

Skinner, Brian J. and Porter, Stephen C., 1995. *The Dynamic Earth: An Introduction to Physical Geology.* John Wiley and Sons, Inc., New York, 567 p.

Tarbuck, Edward J. and Lutgens, Frederick K., 1996. *Earth, an Introduction to Physical Geology,* 5th ed. Prentice Hall, Upper Saddle River, N.J., 605 p.

Specific references for concepts in this chapter and development of the North Cascades geologic story are:

Benioff, H., 1954. Orogenesis and deep crustal structure: additional evidence from seismology. *Geological Society of America Bulletin,* v. 65, pp. 385–400.

Cox, Allan and Hart, Robert Brian, 1986. *Plate Tectonics—How It Works.* Blackwell Scientific Publications, Inc., Boston, 392 p. *Written for brave-hearted geology students with a minor in math, but excellent explanation of the geometry of plate tectonics.*

Glen, William, 1982. *The Road to Jaramillo: Critical Year of the Revolution in Earth Sciences.* Stanford University Press, Stanford, Calif., 459 p. *A detailed history of the plate tectonic revolution and the people who made it happen.*

Kious, W. Jacquelyne and Tilling, Robert L., 1996. *This Dynamic Earth: the Story of Plate Tectonics.* U.S. Geological Survey, Denver, 77 p. *An excellent summary of plate tectonics and its importance. Written for the nongeologist.*

McPhee, John, 1993. *Assembling California.* Farrar, Straus, and Giroux, New York, 304 p. *Wrong state, but a popularized account of building crust with exotic terranes. Good reading, but wanders.*

McPhee, John, 1983. *In Suspect Terrane.* Farrar, Straus, and Giroux, New York, 209 p. *Exotic terranes in the eastern U.S. Good description of how the geologist finds the pieces. Written when the concept was just established in geology.*

Misch, Peter, 1952. Geology of the northern Cascades of Washington. *The Mountaineer,* v. 45, n. 13, pp. 3–22. *First account for laypersons of geology of the North Cascades.*

Misch, Peter, 1966. Tectonic evolution of the northern Cascades of Washington State—a west-cordilleran case history. *A symposium on the tectonic history, mineral deposits of the western Cordillera in British Columbia and in neighboring parts of the U.S.A.* Canadian Institute of Mining and Metallurgy, Vancouver, 1964, Special Volume 8, pp. 101–148. *The first really detailed paper on North Cascades geology, but technical, complex, and labyrinthine.*

Pringle, P. T., 1993. *Roadside Geology of Mount St. Helens National Volcanic Monument and Vicinity.* Washington Department of Natural Resources Division of Geology and Earth Resources Information Circular, v. 88, 120 p.

Raft, A. and Mason, R. G., 1961. Magnetic Survey off the west coast of North America, 40° N. Latitude to 52° N. Latitude. *Geological Society of America Bulletin,* v. 72, n. 8, pp. 1267–70.

Tabor, R. W., 1987. *Geology of Olympic National Park,* 2nd ed. Northwest Interpretive Association, Seattle, 144 p.

CHAPTER 3. RECOGNIZING THE MOUNTAIN MOSAIC

Armstrong, R. L., Harakal, J. E., Brown, E. H., Bernardi, M. L. and Rady, P. M., 1983. Late Paleozoic high-pressure metamorphic rocks in northwestern Washington and southwestern British Columbia: the Vedder Complex. *Geological Society of America Bulletin,* v. 94, pp. 451–458.

Babcock, R. S. and Misch, P., 1988. Evolution of the crystalline core of the North Cascades Range, *in* Ernst, W. G., ed., *Metamorphism and Crustal Evolution of the Western United States* (Rubey Volume VII). Prentice Hall, Englewood Cliffs, N. J., pp. 214–232.

Barksdale, J. D., 1975. Geology of the Methow Valley, Okanogan County, Washington. *Washington Division of Geology and Earth Resources Bulletin,* v. 68, 72 p. *Pioneering work on the Methow Domain.*

Brandon, M. T., 1989. Geology of the San Juan-Cascade Nappes, Northwestern Cascade Range and San Juan Islands, *in* Joseph, N. L., ed., *Geologic guidebook for Washington and adjacent areas.* Washington Division of Geology and Earth Resources Information Circular, v. 86, p. 137–162.

Brown, E. H., 1987. Structural geology and accretionary history of the Northwest Cascade System of Washington and British Columbia. *Geological Society of America Bulletin,* v. 99, p. 201–214.

Brown, E. H., 1988. Metamorphic and structural history of the Northwest Cascades, Washington and British Columbia, *in* Ernst, W. G., ed., *Metamorphism and Crustal Evolution of the Western United States* (Rubey Volume VII). Prentice-Hall, Englewood Cliffs, N. J., pp. 196–213.

Brown, E. H., Blackwell, D. L., Christenson, B. W., Frasse, F. I., Haugerud, R. A., Jones, J. T., Leiggi, P. L., Morrison, M. L., Rady, P. M., Reller, G. J., Sevigny, J. H., Silverberg, D. L., Smith, M. T., Sondergaard, J. N. and Zielger, C. B., 1987. *Geologic Map of the northwest Cascades, Washington*. Geological Society of America Map and Chart Series, MC 61, 10 p. *This is the first real geologic map of a significant part of the North Cascades.*

Brown, E.H. and Gehrels, G.E., 2007. Detrital zircon constraints on terrane ages and affinities and timing of orogenic events in the San Juan Islands and North Cascades, Washington. *Canadian Journal of Earth Science,* v. 44, pp. 1375–1396. *A significant new study. Using a new technique that is leading to revisions of many accepted geologic histories, these results will clarify much of Cordilleran geologic history.*

Davis, G. A., Monger, J. W. H. and Burchfiel, B. C., 1978. Mesozoic construction of the Cordilleran "Collage," central British Columbia to central California, *in* Howell, D. G., and McDougall, K. A., *Mesozoic paleogeography of the western United States*. Pacific Coast Paleogeography Symposium 2: Pacific Section, Society of Economic Paleontologists and Mineralogists, Los Angeles, pp. 1–32. *Historic paper suggesting that parts of the North Cascades were far traveled.*

Haugerud, R. A., 1985. Geology of the Hozameen Group and the Ross Lake shear zone, Maselpanik area, North Cascades, southwest British Columbia. Seattle, University of Washington, Ph.D. thesis, 269 p.

Haugerud, R. A., Brown, E. H., Tabor, R. W., Kriens, B. J. and McGroder, M. F., 1994. Late Cretaceous and early Tertiary orogeny in the North Cascades, *in* Swanson, D. A and Haugerud, R. A., eds., *Geologic Field Trips in the Pacific Northwest*. Published in conjunction with the Geological Society of America Annual Meeting in Seattle, October 24–27 by the Department of Geological Sciences, University of Washington, Seattle, Washington, v. 2, pp. 2E1–2E51.

Kriens, B. and Wernicke, B., 1990. Nature of the contact zone between the North Cascades crystalline core and the Methow sequence in the Ross Lake area, Washington: implications for Cordilleran tectonics. *Tectonics,* v. 9, n. 5, pp. 953–981.

Magloughlin, J. F., 1993. A Nason terrane trilogy. I. Nature and significance of pseudotachylyte; II. Summary of the structural and tectonic history; III. Major and trace element geochemistry, and strontium and neodymium isotope geochemistry of the Chiwaukum Schist, amphibolite, and meta-tonalite gneiss of the Nason terrane. Duluth, University of Minnesota, Ph.D. thesis, 306 p.

Mattinson, J. M., 1972. Ages of zircons from the northern Cascade Mountains, Washington. *Geological Society of America Bulletin,* v. 83, n. 12, pp. 3769–3784. *Landmark study of the ages of North Cascade rocks.*

Matzel, J.E.P. and Bowring, S.A., 2004. Protolith age of the Swakane Gneiss, North Cascades, Washington—Evidence of rapid underthrusting of sediments beneath an arc. *Tectonics,* v. 23, no. TC6009, pp. 1–18.

McGroder, M. F., 1991. Reconciliation of two-sided thrusting, burial metamorphism, and diachronous uplift in the Cascades of Washington and British Columbia. *Geological Society of America Bulletin,* v. 103, n. 2, pp. 189–209.

McGroder, M. F., Garver, J. L. and Mallory, V. S., 1990. Bedrock geologic map, biostratigrapy, and structure sections of the Methow Basin, Washington and British Columbia. Washington Division of Geology and Earth Resources Open File Report, v. 90–19, 32 p.

McTaggart, K. C. and Thompson, R. M., 1967. Geology of part of the northern Cascades in southern British Columbia. *Canadian Journal of Earth Sciences,* v. 4, pp. 1199–1228.

Monger, J. W. H., 1970. Hope map-area, west half, British Columbia. Geological Survey of Canada Department of Energy, Mines and Resources Paper, v. 69–47, 75 p. *Major paper describing rocks of the Chilliwack River terrane.*

Rasbury, T. E. and Walker, N. W., 1992. Implications of Sm-Nd model ages and single grain U-Pb zircon geochronology for the age and heritage of the Swakane Gneiss, Yellow Aster Complex, and Skagit Gneiss, North Cascades, Wash. *Geological Society of America Abstracts with Programs,* v. 24, no. 7, p. A65.

Ray, G. E., 1986. The Hozameen fault system and related Coquihalla serpentine belt of southwestern British Columbia. *Canadian Journal of Earth Science,* v. 23, pp. 1022–1041.

Sevigny, J. H. and Brown, E. H., 1989. Geochemistry and tectonic interpretations of some metavolcanic rock units of the western North Cascades, Washington. *Geological Society of America Bulletin,* v. 101, pp. 391–400.

Sondergaard, J. N., 1979. Stratigraphy and petrology of the Nooksack Group in the Glacier Creek-Skyline divide area, North Cascades, Washington. Bellingham, Western Washington University, M.S. thesis, 103 p.

*Tabor, R. W., Frizzell, V. A., Jr., Whetten, J. T., Waitt, R. B., Jr., Swanson, D. A., Byerly, G. R., Booth, D. B., Hetherington, M. J. and Zartman, R. E., 1987. Geologic map of the Chelan 30-minute by 60-minute quadrangle, Washington. U.S. Geological Survey, Miscellaneous Investigations Map, I-1661.

Tabor, R. W., Zartman, R. E. and Frizzell, V. A., Jr., 1987. Possible tectonostratigraphic terranes in the North Cascades crystalline core, Washington, *in* Schuster, J. E., *Selected Papers on the Geology of Washington*. Washington State Department of Natural Resources Bulletin, v. 77, pp. 107–127.

Tabor, R. W., Haugerud, R. H. and Miller, R. B., 1989. Overview of the geology of the North Cascades, *in* Tabor, R. W., Haugerud, R. H., Brown, E. H., Babcock, R. S. and Miller, R. B., *Accreted Terranes of the North Cascades Range, Washington*. American Geophysical Union, Washington, D.C., pp. 1–33.

*Tabor, R. W., Haugerud, R. A., Brown, E. B. and Hildreth, W., 2003. Geologic Map of the Mount Baker 30-minute by 60-minute quadrangle, Washington. U.S. Geological Survey, Miscellaneous Investigations Map, I-2660. *General geologic map covering much of the area described in this guide.*

*Tabor, R. W., Booth, D. B., Vance, J. A., and Ford, A. B. 2002. Geologic Map of the Sauk River 30- by 60-minute quadrangle, Washington. U.S. Geological Survey Miscellaneous Investigations Map, I-2592. General geologic map covering some of the area described in this guide.

Tennyson, M. E. and Cole, M. R., 1978. Tectonic significance of upper Mesozoic Methow-Pasayten sequence, northeastern Cascade Range, Washington and British Columbia, *in* Howell, D. G. and K. A. McDougall, eds., *Mesozoic Paleogeography of the Western United States*. Pacific Section, Society of Economic Paleontologists and Mineralogists, pp. 499–508.

Trexler, J. H. J. and Bourgeois, J., 1985. Evidence for mid-Cretaceous wrench faulting in the Methow Basin, Washington: tectonostratigraphic setting of the Virginian Ridge Formation. *Tectonics,* v. 4, pp. 379–394.

CHAPTER 4. ANNEALING THE PARTS

Babcock, R. S. and Misch, P., 1989. Origin of the Skagit migmatites, North Cascades Range, Washington State. *Contributions to Mineralogy and Petrology,* v. 101, pp. 485–495.

Haugerud, R. A., van der Heyden, P., Tabor, R. W., Stacey, J. S. and Zartman, R. E., 1991. Late Cretaceous and early Tertiary plutonism and deformation in the Skagit Gneiss Complex, North Cascades Range, Washington and British Columbia. *Geological Society of America Bulletin,* v. 103, pp. 1297–1307.

Monger, J. W. H., Price, R. A. and Tempelman-Kluit, D. J., 1982. Tectonic accretion and the origin of the two major metamorphic and plutonic welts in the Canadian Cordillera. *Geology,* v. 10, n. 2, pp. 70–75.

Whitney, D. L., 1991. Petrogenesis of the Skagit migmatites, North Cascades, Washington: criteria for distinction between anatectic and subsolidus leucosomes in a trondhjemitic migmatite complex. Seattle, University of Washington, Ph.D. thesis, 179 p.

Whitney, D. L., 1992. High-pressure metamorphism in the Western Cordillera of North America: an example from the Skagit Gneiss. *Journal of Metamorphic Geology,* v. 10, pp. 71–85.

Zen, E-an and Hammarstrom, J. M., 1984. Magmatic epidote and its petrologic significance. *Geology,* v. 12, pp. 515–518.

CHAPTER 5. SHIFTING THE PIECES

Beck, M. E., Jr. and Noson, L., 1971. Anomalous paleolatitudes in Cretaceous granitic rocks. *Nature,* Physical Science, v. 235, n. 53, pp. 11–13. *Pioneering study of paleomagnetism indicating that parts of North Cascades had been formed far to the south of where they are now.*

Beck, M. E., Burmester, R. F., Jr. and Schooner, R., 1982. Tertiary paleomagnetism of the North Cascade Range, Washington. *Geophysical Research Letters,* v. 9, n. 5, pp. 515–518.

Miller, R. B., 1994. A mid-crustal contractional stepover zone in a major strike-slip system, North Cascades, Washington. *Journal of Structural Geology,* v. 16, n. 1, pp. 47–60.

Misch, Peter, 1977. Dextral displacements at some major strike faults in the North Cascades (abs.), *in* Geological Association of Canada Cordilleran Section, Vancouver, British Columbia, Programme with Abstracts, v. 2, p. 37.

Vance, J. A. and Miller, R. B., 1981. The movement history of the Straight Creek Fault in Washington State: the last 100 million years (mid Cretaceous to Holocene) (abs). Symposium of geology and mineral deposits in the Canadian Cordillera, Programme with Abstracts, pp. 39–41.

CHAPTER 6. FROM THE FIERY FURNACE

Harris, S. L., 1988. *Fire Mountains of the West.* Mountain Press Publishing Co., Missoula, Mont., 379 p.

*Hildreth, W., 1996. Kulshan caldera: A Quaternary subglacial caldera in the North Cascades, Washington. *Geological Society of America Bulletin,* v. 108, n. 7, pp. 786–793. *Technical but well-written study of a newly recognized and dramatic volcanic event in the Mount Baker area.*

Paterson, S. R., Fowler, Jr., T. K. and Miller, R. B., 1996. Pluton emplacement in arcs: a crustal-scale exchange process. Transactions of the Royal Society of Edinburgh. *Earth Sciences,* v. 87, pp. 115–123.

Tepper, J. H., 1991. Petrology of mafic plutons and their role in granitoid genesis, Chilliwack batholith, North Cascades, Washington. Seattle, University of Washington, Ph.D. thesis, 307 p.

Tepper, J. H., Nelson, B. K., Bergantz, G. W. and Irving, A. J., 1993. Petrology of the Chilliwack batholith, North Cascades, Washington: generation of calc-alkaline granitoids by melting of mafic lower crust with variable water fugacity. *Contributions to Mineralogy and Petrology,* v. 113. pp. 333–351.

Tilling, Robert I., 1987. *Volcanoes.* U.S. Geological Survey General Interest Publication, Denver, 45 p. *A small but excellent booklet explaining volcanoes and their relation to plate tectonics. Available free.*

CHAPTER 7. THE CONSTANT LEVELERS

Booth, D. B., 1987. Timing and processes of deglaciation on the southern part of the Cordilleran Ice Sheet, *in* Ruddiman, W. and Wright, H. O., Jr., eds., North America and Adjacent Oceans During the Last Deglaciation. *The Geology of North America.* Geological Society of America, Boulder, Colorado, v. K-3, pp. 71–90.

*Mackin, J. H. and Cary, A. S., 1965, Origin of Cascade landscapes. Washington Division of Mines and Geology, Information Circular, v. 41, 35 p.

*Manning, H., ed., 1967. *Mountaineering: The Freedom of the Hills,* 2nd ed. The Mountaineers, Seattle, 485 p. *The chapter on Mountain Geology, although out of date for Plate Tectonics, describes many aspects of mountain erosion which are appropriate for the North Cascades. Some drawings from that chapter are reproduced here.*

PART II: GEOLOGIC NOTES FOR POINTS OF INTEREST

Geologic Note 1

Booth *in* Tabor, R. W., Haugerud, R. A., Brown, E. B. and Hildreth, W., 2003, in Chapter 3 reference list.

Concrete Herald, 1984. *North Cascades Traveler's Guide.* Concrete Herald, Concrete, Wash.

Danner, W. R., 1970. Carboniferous system of the western Cordillera of south-western British Columbia and north-western Washington. Compte Rendu 6e Congres Intern. Strat. Geol. Carboniferous. Sheffield, England, 1967, pp. 599–608.

Heller, P. L., 1978. Pleistocene geology and related landslides in the lower Skagit and Baker Valleys, North Cascades, Washington. Bellingham, Western Washington University, M.S. thesis, 154 p.

Geologic Note 3

Riedel, J. L., Pringle, P. T. and Schuster, R.L., 2001, *Deposition of Mount Mazama tephra in landslide-dammed lake on the upper Skagit River, Washington, USA:* Special Publication International Associaton of Sedimentology, v. 30, p. 285–298.

Geologic Note 4

Haugerud and others (1991) in Chapter 4 reference list.

Geologic Note 5

Riedel, J. L. and Haugerud, R. A., 1994. Glacial rearrangement of drainage in the northern North Cascade Range, Washington. Geological Society of America Abstracts with Programs, v. 26, n. 7, p. A-307.

Geologic Note 7

*Pitzer, Paul C., 1978. *Building the Skagit: A Century of Upper Skagit Valley History, 1870–1970.* The Galley Press, Portland, 106 p.

*Jenkins, Will D., *Last Frontier in the North Cascades,* Skagit County Historical Society, 1984. 176 p. *Many interesting tales of early pioneer life along the Skagit River drainage by an author who experienced it. Vivid descriptions of early work on the dams.*

Geologic Note 11

Misch (1966) in Chapter 2 reference list.

Whitney, D. L., and Evans, B. W., 1988. Revised metamorphic history for the Skagit Gneiss, North Cascades: implications for the mechanism of migmatization. Geological Society of America Abstracts with Programs, v. 20, pp. 242–243.

Whitney (1991, 1992) in Chapter 4 reference list.

Geologic Note 12

See Note 7.

Baldwin, J. A., Whitney, D. L. and Hurlow, H. A., 1997. Metamorphic and structural evidence for significant vertical displacement along the Ross Lake fault zone, a major orogen-parallel shear zone in the Cordillera of western North America. *Tectonics,* v. 16, n. 4, pp. 662–681.

Geologic Note 15

Beck, M. E., 1980. Paleomagnetic record of plate-margin tectonic processes along the western edge of North America. *Journal of Geophysical Research,* v. 85, n. B 12, pp. 7115–7131.

Butler, R. F., Gehrels, G. E., McLelland, W. C., May, S. R. and Klepacki, D., 1981. Discordant paleomagnetic poles from the Canadian Coast Plutonic Complex: regional tilt rather than large-scale displacement? *Geology,* v. 17, pp. 691–694.

Geologic Note 18

Beckey, Fred, 1995. *Cascade Alpine Guide: Climbing and High Routes,* 2nd ed., vol. 3. The Mountaineers, Seattle, p. 385.

Geologic Note 20

Snyder, Gary, 1968. *The Back Country.* New Directions Publishing Co., New York.

Geologic Note 22

Tabor, R. W., Haugerud, R. H. and Miller, R. B., 1989, in Chapter 3 reference list.

Geologic Note 31

*Tabor, R. W. and Crowder, D. F., 1968. *Routes and Rocks in the Mount Challenger Quadrangle.* The Mountaineers, Seattle, pp. 34–35. *Out of print, but covers part of this guide. Notes and some figures from this reference have been modified for this book.*

Geologic Note 32

See Note 31, p. 25.

Geologic Note 33

See Note 18, pp. 390–391 and Note 31, pp. 25–26.

Geologic Note 34

See Note 18, pp. 114, 394 and Note 31, pp. 28–29.

Geologic Note 37

See Note 5.

Kerouac, Jack, 1976. *The Dharma Bums.* Penguin Books, New York, p. 243.

Geologic Note 40

Copeland, K. and Copeland, C., 1996. *Don't Waste Your Time in the North Cascades.* Wilderness Press, Berkeley, Calif., 351 p.

Darvill, Jr., Fred T., 1986. *North Cascades Highway Guide,* 2nd ed. Privately published, 63 p.

Geologic Note 47

See Note 1: Heller (1978).

Geologic Note 49

*Hildreth, Wes, *in* Tabor, R. W., Haugerud, R. A., Brown, E. B. and Hildreth, W., 2003, in Chapter 3 reference list. *Description of many volcanic features associated with Mount Baker.*

Geologic Note 51

Tabor (1987), p. 89, in Chapter 2 reference list.

Geologic Note 52

See Note 31, pp. 22–23 and Note 18, p. 384.

Geologic Note 53

See Note 49.

Geologic Note 54

Babcock, S., 1996. Ancient aires and rock romancing, *in* Miles, J. C., ed., *Impressions of the North Cascades.* The Mountaineers, Seattle, pp. 15–28.

Misch (1966) in Chapter 2 reference list.

Rasbury, T. E. and Walker, N. W. (1992) in Chapter 3 reference list.

Geologic Note 58

Armstrong and others (1983) in Chapter 3 reference list.

Geologic Note 59

*Booth, D. B. and Goldstein, B., 1994. Patterns and processes of landscape development by the Puget lobe ice sheet, *in* Cheney, E. S. and Lasmanis, R., eds., *Regional Geology of Washington State,* Washington Division of Geology and Earth Resources Bulletin, v. 80, pp. 207–218.

Geologic Note 63

Huntting, M. T., 1956. *Inventory of Washington Minerals,* Part II, Metallic Minerals. Washington Division of Mines and Geology Bulletin 37, 428 p.

Geologic Note 64

Tabor, R. W., 1963, Large quartz diorite dike and associated explosion breccia, northern Cascade Mountains, Washington. *Geological Society of America Bulletin,* v. 74, pp. 1203–1208.

Geologic Notes 65 and 66

Brown, E. H., Cary, J. A., Dougan, B. E., Dragovich, J. D., Fluke, S. M. and McShane, D. P., 1994. Tectonic evolution of the Cascades crystalline core in the Cascade River area, Washington, *in* Cheney, E. S. and Lasmanis, R., eds., *Regional Geology of Washington State.* Washington Division of Geology and Earth Resources Bulletin, v. 80, pp. 93–113.

Geologic Note 67

Dragovich, J. D., 1989. Petrology and structure of the Cascade River Schist in the Sibley Creek area, Northern Cascades, Washington. Bellingham, Western Washington University, M.S. thesis, 167 p.

Geologic Note 68

Tabor and others (1987) in Chapter 3 reference list.

Geologic Note 72

Tabor, R. W., 1961. The crystalline geology of the area south of Cascade Pass, northern Cascade Mountains, Washington. Seattle, University of Washington, Ph.D. thesis, 205 p.

Dougan, B. E., 1993. Structure and metamorphism in the Magic Mountain–Johannesburg Mountain area, North Cascades, Washington. Bellingham, Western Washington University, M.S. thesis, 110 p.

Dougan, B. E. and Brown, E. H., 1991. Structure and metamorphism in the Magic Mtn.–Johannesburg Mtn. area, north Cascades, Washington. Geological Society of America Abstracts with Programs, v. 23. n. 2, p. 19.

*Tabor, R. W., Booth, D. B. and Ford, A. B., 2002. in Chapter 3 reference list. *Summarizes arguments on the origin of the Magic Mountain Gneiss.*

Geologic Note 73

Carpenter, M. R., 1993, The Church Mountain sturzstrom (mega-landslide) near Glacier, Washington. Bellingham, Western Washington University, M.S. thesis, 71 p.

Carpenter, M. R., and Easterbrook, D. J., 1993. The Church Mountain sturzstrom (mega-landslide), Glacier, Washington (abs.). Geological Society of America Abstracts with Programs, v. 25, n. 5, p. 18.

Cary, C. M., Easterbrook, D. J. and Carpenter, M. R., 1992. Post-glacial mega-landslides in the North Cascades near Mt. Baker, Washington. *Geological Society of America Abstracts with Programs,* v. 24, n. 5, p. 13.

Geologic Note 75

See Note 49.

Geologic Note 76

Tabor, R. W., Haugerud, R. A., Brown, E. B. and Hildreth, W., (2003) in Chapter 3 reference list.

Geologic Note 77

Hildreth, W. (1996) in Chapter 6 reference list.

Geologic Note 79

Hildreth, W. (1996) in Chapter 6 reference list.

Geologic Note 80

Tabor, R. W., Haugerud, R. A., Brown, E. B. and Hildreth, W., 2003, in Chapter 3 reference list.

See Note 77.

Geologic Note 81

Brown, E. H., 1986. Geology of the Shuksan suite, North Cascades, Washington, U.S.A., *in* Evans, B. W., and Brown, E. H., eds., *Blueschists and Eclogites.* Geological Society of America Memoir 164, pp. 143–153.

Geologic Note 82

Wes Hildreth supplied the photograph on which figure 78 is based.

Geologic Note 84

See Note 49.

Geologic Note 86

See Note 77.

Geologic Note 87

Misch (1966) in Chapter 2 reference list.

Brown and others (1987) in Chapter 3 reference list.

Sevigny, J. H. and Brown, E. H. (1989) in Chapter 3 reference list.

Geologic Note 90

Moen, W. S., 1969. Mines and mineral deposits of Whatcom County, Wa. Washington Division of Mines and Geology Bulletin 57, 133 p.

Geologic Note 93

Tepper, J. H. (1996) in Chapter 6 reference list.

Geologic Note 95

Tepper, J. H. (1996) in Chapter 6 reference list.

Geologic Note 96

See Note 18, p. 60 and Note 31, p. 20.

Geologic Note 97

See Note 5.

Geologic Note 104

Haugerud, R. A. (1985) in Chapter 3 reference list.

Geologic Note 107

*Crowder, D. F and Tabor, R. W., 1965, *Routes and Rocks: Hiker's Guide to the North Cascades from Glacier Peak to Lake Chelan:* The Mountaineers, Seattle, 235 p. *Long out of print, but covers some of the area described in this guide. Many of the Geologic Notes and figures from this reference have been used in this guide, with modest to considerable modification.*

Geologic Note 108

Magloughlin, J. F. (1993) in Chapter 3 reference list.

See Note 72: Tabor, R. W., Booth, D. B. and Ford, A. B., 2002.

Geologic Note 110

See Note 107, p. 20.

Geologic Note 111

See Note 31, p. 23.

*Crowder, D. F., Tabor, R. W. and Ford, A. B., 1966. Geologic map of the Glacier Peak quadrangle, Snohomish and Chelan Counties, Washington. U.S. Geological Survey Geologic Quadrangle Map, GQ 473. *Detailed geologic map of southernmost part of this guide.*

Geologic Note 112
See Note 107, p. 39.
Geologic Note 113
See Note 107, pp. 40–41.
Geologic Note 114
See Note 107, pp. 42–43.
Geologic Note 115
See Note 107, p. 47, 52.
*McPhee, John A., 1971. *Encounters with the Archdruid.* Farrar, Straus, and Giroux, New York, 245 p. *Lively discussion between Dave Brower (the Archdruid) and Professor Charles Park (mining geologist) at the site of the Glacier Peak Mine.*
Geologic Note 116
See Note 107, p. 50.
Geologic Note 117
See Note 107, pp. 169, 186.
Whetten, J. T., 1967. Lake Chelan, Washington: bottom and sub-bottom topography. *Limnology and Oceanography,* v. 12, n. 2, pp. 253–259.
Geologic Note 127
*Britt, D., 1893. The Heart of the Cascades, *in* Stone, C., compiler, *Stehekin: Glimpses of the Past, a Collection of Early Writings.* Privately published, 1983.
Geologic Note 129
Tabor, R. W., Haugerud, R. A., Brown, E. H., Babcock, R. S., and Miller, R. B. (1989) in Chapter 3 reference list.
Geologic Note 130
See Note 107, p. 186.
*Cater, F. W. and Crowder, D. F., 1967. Geologic map of the Holden quadrangle, Snohomish and Chelan Counties, Washington. U.S. Geological Survey, Geologic Quadrangle Map, GQ 646. *Detailed geologic map covers southeasternmost part of this guide.*
Dragovich, J. D. and Derkey, R. E., 1994. A Late Triassic island-arc setting for the Holden volcanogenic massive sulfide deposit, North Cascades, Washington. *Washington Geology,* v. 22, n. 1, pp. 28–39.
Geologic Note 131
See Note 107, p. 195.
Geologic Note 132
See Note 107, p. 195.
Geologic Note 133
See Note 107, p. 196.
Geologic Note 134
See Note 107, p. 198–199.
Geologic Note 135
See Note 107, pp. 197, 198.
Geologic Note 136
See Note 107, p. 194.
Beckey, F., 1989. *Cascade Alpine Guide: Climbing and High Routes,* vol. 2. The Mountaineers, Seattle, pp. 208–209.
Geologic Note 138
Johannsen, Albert, 1932. *A Descriptive Petrography of the Igneous Rocks,* vol. II. University of Chicago Press, pp. 243–244.
Geologic Note 140
Barksdale, J. D. (1975) in Chapter 3 reference list.
Geologic Note 141
Roe, JoAnn, 1997. *North Cascades Highway.* The Mountaineers, Seattle, 175 p.
Geologic Note 142
See Note 140.
Haugerud, R. A., and others (1994) in Chapter 3 reference list.
*Haugerud, R. A. and Tabor, R. W., 2009. Geologic Map of the North Cascade Range, Washington. U.S. Geological Survey Scientific Investigations Map 2940, scale 1:200,000 [http://pubs.usgs.gov/sim/2940/]. *A regional geologic map of the North Cascades from south of Snoqualmie Pass to the Canadian border, including a non-technical text and a CD with about 100 photos of mountains and rocks.*
Geologic Note 143
See Note 90 and Note 141: Roe (1997).
Staatz, M. H., Weiss, P. L., Tabor, R. W., Robertson, J. F., Van Noy, R. M., Pattee, E. C. and Holt, D. C., 1971. *Mineral Resources of the Pasayten Wilderness Area, Washington.* U.S. Geological Survey Bulletin 1325, 255 p.
Geologic Note 145
See Note 143 and Note 90.
Geologic Note 146
Barksdale, J. D. (1975) in Chapter 3 reference list.
See Note 142: Haugerud, R. A. and Tabor, R. W. (2009).
Geologic Note 148
See Note 143.
Geologic Note 150
See Note 142: Haugerud, R. A. and Tabor, R. W. (2009).
Geologic Note 151
See Note 142: Haugerud, R. A. and Tabor, R. W. (2009).
Geologic Note 152
Waitt, R. B., Jr., 1979. Rockslide-avalanche across distributary of Cordilleran ice in Pasayten Valley, Northern Washington. *Arctic and Alpine Research,* v. 11, n. 1, pp. 33–40.
Geologic Note 153
See Note 142: Haugerud, R. A. and Tabor, R. W. (2009).
Geologic Note 154
See Note 142: Haugerud, R. A. and Tabor, R. W. (2009).

GEOLOGIC MAP, PLATE 2

Cater, F. W. and Crowder, D. F., 1967. Geologic map of the Holden quadrangle, Snohomish and Chelan Counties, Washington. U.S. Geological Survey, GQ 646.
Cater, F. W. and Wright, T. L., 1967. Geologic map of the Lucerne quadrangle, Chelan County, Washington. U.S. Geological Survey, GQ 647.
Crowder, D. F., Tabor, R. W. and Ford, A. B., 1966. Geologic map of the Glacier Peak quadrangle, Snohomish and Chelan Counties, Washington. U.S. Geological Survey, GQ 473.
Dragovich, J. D. and Norman, D. K., 1995. Geologic map of the west half of the Twisp 1:100,000 quadrangle, Washington. Washington Division of Geology and Earth Resources, Open File Report, 95-3, 63 p.
*Haugerud, R. A. and Tabor, R. W., 2009. Geologic Map of the North Cascade Range, Washington. U.S. Geological Survey Scientific Investigations Map 2940, scale 1:200,000.
Miller, R. B., 1987. Geologic map of the Twisp River-Chelan divide region, Washington. Washington Division of Geology and Earth Resources, Open File Report 87-17.
Monger, J. W. H., 1989. Geology, Hope, British Columbia, Map 41-1989.
Tabor, R. W., 1961. The crystalline geology of the area south of Cascade Pass, northern Cascade Mountains, Washington. Seattle, University of Washington, Ph.D. thesis, 205 p.
Tabor, R. W., Booth, D. B. and Ford, A. B., 2002. Geologic map of the Sauk River 30-by-60-minute quadrangle. U. S. Geological Survey, I-2592
Tabor, R. W., Haugerud, R. A., Hildreth, W. and Brown, E. H., 2003. Geologic map of the Mount Baker 30-by-60-minute quadrangle, Washington. U.S. Geological Survey, I-2660.

Index

THE MOUNTAINEERS, founded in 1906, is a nonprofit outdoor activity and conservation organization, whose mission is "to explore, study, preserve, and enjoy the natural beauty of the outdoors. . . ." Based in Seattle, Washington, it is now one of the largest such organizations in the United States, with seven branches throughout Washington State.

The Mountaineers sponsors both classes and year-round outdoor activities in the Pacific Northwest, which include hiking, mountain climbing, ski-touring, snowshoeing, bicycling, camping, kayaking and canoeing, nature study, sailing, and adventure travel. The organization's conservation division supports environmental causes through educational activities, sponsoring legislation, and presenting informational programs. All its activities are led by skilled, experienced volunteers, who are dedicated to promoting safe and responsible enjoyment and preservation of the outdoors.

If you would like to participate in these organized outdoor activities or the organization's programs, consider a membership in The Mountaineers. For information and an application, write or call The Mountaineers, 7700 Sand Point Way NE, Seattle, Washington 98115; (206) 521-6001.

The Mountaineers Books, an active, nonprofit publishing program of the organization, produces guidebooks, instructional texts, historical works, natural history guides, and works on environmental conservation. All books produced by The Mountaineers Books are aimed at fulfilling the organization's mission.